UNIVERSITY OF S
30125 00799818 9

High Pressure Processing of Foods

The *IFT Press* series reflects the mission of the Institute of Food Technologists—advancing the science and technology of food through the exchange of knowledge. Developed in partnership with Blackwell Publishing, *IFT Press* books serve as leading edge handbooks for industrial application and reference and as essential texts for academic programs. Crafted through rigorous peer review and meticulous research, *IFT Press* publications represent the latest, most significant resources available to food scientists and related agriculture professionals worldwide.

High Pressure Processing of Foods

EDITORS

Christopher J. Doona • Florence E. Feeherry

FOREWORD BY

C. Patrick Dunne

Christopher J. Doona, PhD, serves with distinction as Research Chemist of the Department of Defense (DoD) Combat Feeding Directorate at the US Army Natick Soldier Research, Development, & Engineering Center, with responsibilities for independently and in collaboration with other research scientists conducting, coordinating, and executing new areas of basic and applied research with potential benefits to the military. Dr. Doona's research investigations specialize in novel chemical heating and disinfection technologies, and novel food processing technologies such as ohmic heating, microwaves, and high pressure processing (HPP) for the safety and stability of foods.

Florence E. Feeherry, MS, serves with distinction as Research Microbiologist for the DoD Combat Feeding Directorate at NSRDEC, with responsibilities for independently and in collaboration with other researchers carrying out basic and applied research investigations specializing in the principles of food microbiology to establish the safety of foods stabilized by hurdles or processed with novel technologies such as HPP.

Blackwell Publishing Professional
2121 State Avenue, Ames, Iowa 50014, USA

Orders: 1-800-862-6657
Office: 1-515-292-0140
Fax: 1-515-292-3348
Web site: www.blackwellprofessional.com

Blackwell Publishing Ltd
9600 Garsington Road, Oxford OX4 2DQ, UK
Tel.: +44 (0)1865 776868

Blackwell Publishing Asia
550 Swanston Street, Carlton, Victoria 3053, Australia
Tel.: +61 (0)3 8359 1011

First edition, 2007

Library of Congress Cataloging-in-Publication Data
High pressure processing of foods / editors, Christopher J. Doona, Florence E. Feeherry.
— 1st ed.
p. cm.
Includes bibliographical references and index.
ISBN 978-0-8138-0944-1 (alk. paper)
1. Food industry and trade. 2. High pressure (Technology) I. Doona, Christopher J.
II. Feeherry, Florence E.
TP371.75.H54 2007 664′.02—dc22
2007019842

The last digit is the print number: 9 8 7 6 5 4 3 2 1

We thank our families for their unwavering support in this, as in all, endeavors. To Peter Feeherry for humor and good judgment, Helen Murphy Doona for spirit and determination, we express our gratitude, and to Darleen and Murphy Doona the heartfelt gratitude of love that grows more boundless every day. To Irwin Taub, in memoriam,

We also remember an old friend,
who reminds us to always forge ahead;
Persistence & integrity are key,
with challenge comes opportunity.

Titles in the *IFT Press* series

- *Accelerating New Food Product Design and Development* (Jacqueline H. Beckley, Elizabeth J. Topp, M. Michele Foley, J.C. Huang, and Witoon Prinyawiwatkul)
- *Biofilms in the Food Environment* (Hans P. Blaschek, Hua Wang, and Meredith E. Agle)
- *Calorimetry and Food Process Design* (Gönül Kaletunç)
- *Food Ingredients for the Global Market* (Yao-Wen Huang and Claire L. Kruger)
- *Food Irradiation Research and Technology* (Christopher H. Sommers and Xuetong Fan)
- *Food Risk and Crisis Communication* (Anthony O. Flood and Christine M. Bruhn)
- *Foodborne Pathogens in the Food Processing Environment: Sources, Detection and Control* (Sadhana Ravishankar and Vijay K. Juneja)
- *High Pressure Processing of Foods* (Christopher J. Doona and Florence E. Feeherry)
- *Hydrocolloids in Food Processing* (Thomas R. Laaman)
- *Microbiology and Technology of Fermented Foods* (Robert W. Hutkins)
- *Multivariate and Probabilistic Analyses of Sensory Science Problems* (Jean-Francois Meullenet, Rui Xiong, and Chris Findlay)
- *Nondestructive Testing of Food Quality* (Joseph Irudayaraj and Christoph Reh)
- *Nonthermal Processing Technologies for Food* (Howard Q. Zhang, Gustavo V. Barbosa-Canovas, V.M. Balasubramaniam, Editors; C. Patrick Dunne, Daniel F. Farkas, James T.C. Yuan, Associate Editors)
- *Nutraceuticals, Glycemic Health and Diabetes* (Vijai K. Pasupuleti and James W. Anderson)
- *Packaging for Nonthermal Processing of Food* (J. H. Han)
- *Preharvest and Postharvest Food Safety: Contemporary Issues and Future Directions* (Ross C. Beier, Suresh D. Pillai, and Timothy D. Phillips, Editors; Richard L. Ziprin, Associate Editor)
- *Processing and Nutrition of Fats and Oils* (Ernesto M. Hernandez, Monjur Hossen, and Afaf Kamal-Eldin)
- *Regulation of Functional Foods and Nutraceuticals: A Global Perspective* (Clare M. Hasler)
- *Sensory and Consumer Research in Food Product Design and Development* (Howard R. Moskowitz, Jacqueline H. Beckley, and Anna V.A. Resurreccion)
- *Thermal Processing of Foods: Control and Automation* (K.P. Sandeep)
- *Water Activity in Foods: Fundamentals and Applications* (Gustavo V. Barbosa-Canovas, Anthony J. Fontana Jr., Shelly J. Schmidt, and Theodore P. Labuza)
- *Whey Processing, Functionality and Health Benefits* (Charles I. Onwulata and Peter J. Huth)

CONTENTS

CONTRIBUTORS

Abram Aertsen (chapter 4)
Katholieke Universiteit Leuven, Department of Microbial and Molecular Systems, Laboratory of Food Microbiology, Leuven, Belgium

Murad A. Al-Holy (corresponding author, chapter 3)
Hashemite University, Department of Clinical Nutrition and Dietetics, Faculty of Allied Health Sciences, Zarqa, Jordan. Phone: +11-962-390-3333, Fax: +11-962- 538-26613, E-mail: murad@hu.edu.jo

Rick Bell (chapter 10)
Product Optimization and Evaluation Team, US Army Natick Soldier Research, Development, & Engineering Center, Natick, MA 01760

Stanley Brul (corresponding author, chapter 8)
University of Amsterdam, Department of Molecular Biology & Microbial Food Safety, Swammerdam Institute for Life Sciences, The Netherlands, Department of Advanced Food Microbiology, Unilever Food and Health Research Institute, Olivier van Noortlaan 120, 3133 AT Vlaardingen, The Netherlands, Phone: 31-10-4605161, Fax: 31-10-4605188, E-mail: stanley.brul@unilever.com; brul@science.uva.nl

Roman Buckow (chapter 5)
Technische Universität Berlin, Fachgebiet Lebensmittelbiotechnologie und–prozesstechnik, Königin Luise Str. 22, D-14195, Berlin, Germany

Armand V. Cardello (chapter 10)
Product Optimization and Evaluation Team, US Army Natick Soldier Research, Development, & Engineering Center, Natick, MA 01760

Yoon-Kyung Chung (chapter 7)
The Ohio State University, Department of Food Science and Technology, Parker Food Science Building, 2015 Fyffe Road, Columbus, OH 43210

Maria Corradini (chapter 6)
University of Massachusetts-Amherst, Department of Food Science, Chenoweth Laboratory, Amherst, MA 01003

Christopher J. Doona (chapter 1 and corresponding author, chapter 6)
DoD Combat Feeding Directorate, US Army Natick Soldier Research, Development, & Engineering Center, Natick, MA 01760

C. Patrick Dunne (Foreword)
DoD Combat Feeding Directorate, US Army Natick Soldier Research, Development, & Engineering Center, Natick, MA 01760

Matthias A. Ehrmann (chapter 5)
Technische Universität München, Lehrstuhl Technische Mikrobiologie, Weihenstephaner Steig 16, D-85350, Freising-Weihenstephan, Germany

Florence E. Feeherry (chapter 6)
DoD Combat Feeding Directorate, US Army Natick Soldier Research, Development, & Engineering Center, Natick, MA 01760

Michael G. Gänzle (corresponding author, chapter 5)
University of Alberta, Department of Agricultural, Food and Nutritional Science, 4-10 Ag / For Centre, Edmonton, Alberta, Canada T6G 2P5, Phone: + 1 789 492 0774, Fax: + 1 780 492 4265, E-mail: michael.gaenzle@ualberta.ca

Volker Heinz (chapter 5)
Technische Universität Berlin, Fachgebiet Lebensmittelbiotechnologie und–prozesstechnik, Königin Luise Str. 22, D-14195, Berlin, Germany, current address: German Institute of Food Technology (DIL e.V.), Quakenbrück, Germany

Bart J.F. Keijser (chapter 8)
University of Amsterdam, Department of Molecular Biology & Microbial Food Safety, Swammerdam Institute for Life Sciences, The Netherlands, current address: TNO Quality of Life, Department of Microbiology, Zeist, The Netherlands

Ming H. Lau (corresponding author, chapter 9)
Kraft Foods Global, Inc., Strategic Research, 801 Waukegan Rd., Glenview, IL 60025

Mengshi Lin (chapter 3)
Washington State University, Department of Food Science and Human Nutrition, Box 646376, Pullman, WA 99164-6373

Mark Linton (chapter 1)
Agri-food and Biosciences Institute, Agriculture, Food and Environmental Science Division (Food Microbiology Branch), Newforge Lane, Belfast, United Kingdom BT9 5PX

Aaron S. Malone (chapter 7)
The Ohio State University, Department of Food Science and Technology, Parker Food Science Building, 2015 Fyffe Road, Columbus, Ohio 43210

Dirk Margosch (chapter 5)
Technische Universität München, Lehrstuhl Technische Mikrobiologie, Weihenstephaner Steig 16, D-85350, Freising-Weihenstephan, Germany

Chris W. Michiels (corresponding author, chapter 4)
Katholieke Universiteit Leuven, Department of Microbial and Molecular Systems, Laboratory of Food Microbiology, Kasteelpark Arenberg 22, B-3001 Leuven, Belgium, Phone: +32-16-321578, Fax: +32-16-321960, E-mail: chris.michiels@biw.kuleuven.be

Roy C. Montijn (chapter 8)
TNO Quality of Life, Department of Microbiology, Zeist, The Netherlands

Suus J.C.M. Oomes (chapter 8)
Department of Advanced Food Microbiology, Unilever Food and Health Research Institute, Olivier van Noortlaan 120, 3133 AT Vlaardingen, The Netherlands

Margaret F. Patterson (corresponding author, chapter 1)
Agri-food and Biosciences Institute, Agriculture, Food and Environmental Science Division (Food Microbiology Branch), Newforge Lane, Belfast, United Kingdom BT9 5PX, and Department of Food Science, Queen's University Belfast, Newforge Lane, Belfast, United Kingdom BT9 5PX

Micha Peleg (chapter 6)
University of Massachusetts-Amherst, Department of Food Science, Chenoweth Laboratory, Amherst, MA 01003

Barbara A. Rasco (chapter 3)
Washington State University, Department of Food Science and Human Nutrition, Box 646376, Pullman, WA 99164-6373

Edward W. Ross (chapter 6)
DoD Combat Feeding Directorate, US Army Natick Soldier Research, Development, & Engineering Center, Natick, MA 01760

Frank H.J. Schuren (chapter 8)
TNO Quality of Life, Department of Microbiology, Zeist, The Netherlands

Peter Setlow (corresponding author, chapter 2)
University of Connecticut Health Center, Department of Molecular, Microbial and Structural Biology, Farmington, CT 06030-3305, Phone: 860-679-2607, Fax: 860-679-3408, E-mail: setlow@nso2.uchc.edu

Edmund Ting (corresponding author, chapter 11)
Pressure BioScience, Inc., 321 Manley Rd., West Bridgewater, MA 02379

Evan J. Turek (chapter 9)
Kraft Foods Global, Inc., Strategic Research, 801 Waukegan Rd., Glenview, IL 60025

Hans van der Spek (chapter 8)
University of Amsterdam, Department of Molecular Biology & Microbial Food Safety, Swammerdam Institute for Life Sciences, The Netherlands

Isabelle Van Opstal (chapter 4)
Katholieke Universiteit Leuven, Department of Microbial and Molecular Systems, Laboratory of Food Microbiology, Leuven, Belgium

Rudi F. Vogel (chapter 5)
Technische Universität München, Lehrstuhl Technische Mikrobiologie, Weihenstephaner Steig 16, D-85350, Freising-Weihenstephan, Germany

Alan O. Wright (corresponding author, chapter 10)
Product Optimization and Evaluation Team, US Army Natick Soldier Research, Development, & Engineering Center, Natick, MA 01760

Ahmed E. Yousef (corresponding author, chapter 7)
The Ohio State University, Department of Food Science and Technology, Parker Food Science Building, 2015 Fyffe Road, Columbus, Ohio 43210, Phone: 614-292-7814, Fax: 614-292-0218, E-mail: yousef.1@osu.edu

FOREWORD

It was with extreme pleasure and appreciation that I accepted the invitation of editors Christopher Doona and Florence "Chickie" Feeherry to write a foreword for *High Pressure Processing of Foods*. All scientists and engineers interested in nonthermal processing technologies will certainly find *High Pressure Processing of Foods* useful and timely, and I am grateful to Chris and Chickie for their dedication in bringing this book to life as an outstanding resource of lasting value, and for their accomplishments in HPP research at the Department of Defense (DoD) Combat Feeding Directorate at the US Army Natick Soldier Research, Development, & Engineering Center (NSRDEC), which also include a collaborative chapter in this impressive volume.

The modern inception of HPP for food processing began in the late 1980s at the University of Delaware with a meeting of great minds: the visionary (Professor and Food Engineer Dan Farkas), the microbiologist (Professor Dallas Hoover), and the well-rounded expert (Professor Dietrich Knorr). Dan Farkas presented a seminar at NSRDEC in 1988 in relation to the use of high pressure for undersea depots for the Deep-Sea Forward program. The idea of HPP as a food processing technology immediately caught my attention as a biochemist. In 1991, NSRDEC provided funding to begin an important research initiative to concomitantly improve military rations and commercial foods by HPP. This initiative involved Dan Farkas (who had moved to Oregon State University) and Dallas Hoover and led to their landmark paper entitled simply "High Pressure Processing," which appeared in the 2000 supplement to the *Journal of Food Science* (pp. 47–64).

With successful research advances in this program leading the way, NSRDEC assumed greater interest in the development of HPP and provided funding for follow-on projects with my stewardship. Under a Broad Agency Announcement, Professor Bibek Ray (University of

Wyoming) carried out fundamental work on the synergies of HPP and bacteriocins (anti-microbial peptides used as food additives) for the preservation of meats that led to several related publications. Professor Hoover, Dr. Cindy Stewart, myself, and Dr. Anthony Sikes (NSRDEC) collaborated to produce US Patent 6,110,516, entitled "Process for treating foods using saccharide esters and superatmospheric hydrostatic pressure."

I am thankful to the DoD Combat Feeding Directorate – NSRDEC for the uniquely rich opportunity of becoming involved with HPP of foods that has given me the privilege and good fortune of working with a large assembly of talented people from many disciplines for the past 16 years, people whose dedicated efforts have helped move this technology forward from its nascent stages to commercialization. I am especially thankful to the many distinguished scientists and engineers in nonthermal processing for their recognition of my contributions as co-founder (along with Huub Lelieveld of Unilever) of the Institute of Food Technologists' (IFT's) Nonthermal Processing Division and for choosing me to serve as its first chair in 2000. My appreciation extends to a wide network of colleagues who supported the 2005 IFT Myron Solberg Award I received, honoring an IFT member for providing leadership and excellence in the establishment, successful development, and continuation of an industry/government/academia cooperative organization. In 2005, Dr. Edmund Ting of Avure Technologies (formerly part of Flow International) and I received a Federal Laboratory Consortium Award for Excellence in Technology Transfer for High-Pressure Food Processing to Provide Increased Safety and Quality. The basis for this award was the commercialization of HPP that was carried out under the auspices of the Army's Dual-Use Science and Technology program dedicated to simultaneously meeting the demands of the consumer marketplace and the military for expanding variety and improving the quality of shelf-stable combat rations for food products containing whole muscle meats, eggs, potato, and pasta products. I am thankful to the Army for their appreciation bestowed through numerous awards and citations for my involvement in efforts striving to improve the variety, quality, and nutritive value of operational military rations through the uses of HPP.

While the possible applications of HPP are only just beginning to be realized, nonthermal processing of foods in general, and HPP in particular, will remain an exciting and inspiring field of food science research that will surely be emphasized in finding solutions to new challenges in food preservation and safety in the increasingly global, complex supply chain and marketplace for the twenty-first century and in the production of fresher, safe foods for all consumers to enjoy.

C. Patrick Dunne
Senior Science Advisor
DoD Combat Feeding Directorate
US Army Natick Soldier Research
Development, & Engineering Center

PROLOGUE

I do not mean to say we are bound to follow implicitly in whatever our fathers did. To do so, would be to discard all the lights of current experience—to reject all progress—all improvement. What I do say is, that if we would supplant the opinions and policy of our fathers in any case, we should do so upon evidence so conclusive, and argument so clear, that even their great authority, fairly considered and weighed, cannot stand.

Abraham Lincoln
Cooper Union Address
February 27, 1860
New York

PREFACE

High pressure processing (HPP) is a leading nonthermal food processing technology that is often cited as a major technological innovation in food preservation. While it is too early to assure its place in food preservation history among breakthroughs such as Appert's discovery of canning, Pasteur and Bernard's use of heat to kill foodborne microorganisms, or Birdseye's development of frozen foods, HPP has emerged as a viable commercial alternative for the pasteurization of value-added fruits, vegetables, meat, and seafood products that are safely enjoyed by today's consumer. HPP also has the capacity to inactivate *Clostridium botulinum* and other bacterial spores, and the food industry, government agencies, and academia are intensifying their efforts to develop HPP methods for inactivating foodborne bacterial spores in foods. Such products will feature more fresh-like character and improved quality attributes for the consumer and might well be expected to significantly impact the multi-billion-dollar market for low-acid, shelf-stable canned foods. *High Pressure Processing of Foods* is intended to capture the current state of scientific knowledge regarding the use of HPP to inactivate bacterial spores as a starting point for future research that will lead to the development of commercially sterile low-acid foods.

High Pressure Processing of Foods culminates to date the scientific advances of leading experts in academia, industry, and government agencies exploring microbial inactivation for the safe preservation of foods by HPP. Patterson et al. (chapter 1) provide a historical introduction to HPP; then in a broad sense, Setlow (chapter 2), Al-Holy et al. (chapter 3), and Van Opstal et al. (chapter 4) use molecular techniques to explore various aspects of the mechanisms of spore germination and inactivation; and Gänzle et al. (chapter 5) and Chung et al. (chapter 7) cover the influence of various processing parameters (e.g., high pressure, temperature, food matrix properties, and the presence of anti-microbial or sensitizing compounds) on microbial inactivation. Since the perspective of the consumer is critical for the commercial success of HPP-treated foods, Lau and Turek (from Kraft Global Foods; chapter 9) and Wright et al. (chapter 10) incorporate consumer analysis of HPP-treated foods in their chapters. Brul et al. (Unilever Food and Health Research Institute; chapter 8) present future models for food quality assurance systems. Doona et al. (chapter 6) use nonlinear mathematical models to characterize and predict microbial inactivation, and Ting (chapter 11) discusses controlling temperature with HPP equipment. It has been our pleasure and privilege to work with this collection of esteemed scientists.

HPP is an important avenue of present and future research at DoD Combat Feeding Directorate – US Army Natick Soldier Research, Development, & Engineering Center (NSRDEC). Our interest in HPP originated from the application of pressure in inorganic reaction kinetics, which later melded with research involving intrinsic chemical markers and pathogen modeling. Presently, our HPP research involves the application of the Quasi-chemical kinetics model for a more complete understanding of bacterial spore inactivation. From these interests, we conceived a progression of high-profile, cutting-edge symposia at IFT annual meetings (co-sponsored by the Nonthermal Processing Division and the Food Microbiology Division) that formed the foundation for *High Pressure Processing of Foods*.

What better place to have begun this journey than Chicago, with the first symposium, "Science-Based Applications of High Pressure Processing in the Food Industry," so near in proximity to the stockyards and meat packing companies that were described in Upton Sinclair's *The Jungle*? Publication of *The Jungle* led to President Theodore Roosevelt's support of the 1906 Pure Food and Drug Act and Meat Inspection Act to control the safety of the nation's food supply. The second symposium, "Inactivating Pathogens, Parasites, and Viruses Using High Pressure Processing and Other Emerging Technologies," in Las Vegas broadened the knowledge base to include other types of organisms and additional alternative food processing technologies. The third symposium, "Mechanisms and Modeling of Bacterial Spore Inactivation by High Pressure Processing," and held in pre-Katrina New Orleans, addressed the inactivation of bacterial spores and provided the primary driving force for the development of *High Pressure Processing of Foods*.

With Lincoln's rationale, as quoted above, serving as a metaphor describing scientific advancement, let us assert that *High Pressure Processing of Foods* conveys the experiences of esteemed scientists who demonstrate they were not bound simply to follow implicitly in their pursuit of scientific knowledge, but set out on bold new paths of scientific exploration and innovation, with their progress lighting the way toward the commercialization of shelf-stable foods. It is our hope that research efforts such as these shall provide a basis for future scientific evidence so conclusive and argument so clear that sometime soon—in the beginning of the second century since publication of *The Jungle* and passage of the 1906 Pure Food and Drug Act—the use of HPP will become viable for the commercial sterilization of high quality, safe, low-acid foods.

Christopher J. Doona
Florence E. Feeherry

ACKNOWLEDGMENTS

We would like to acknowledge with gratitude and appreciation NSRDEC – DoD CFD Director Gerald Darsch for his support, Professor Moo-Yeol Baik of Kyung-Hee University, South Korea, for stimulating our interests in high pressure research, and Professor Mary Ellen Doona for providing insightful comments in reading parts of this manuscript.

Chapter 1

Introduction to High Pressure Processing of Foods

Margaret F. Patterson, Mark Linton, and Christopher J. Doona

Introduction

The modern consumer requires foods that are safe and nutritious, free from additives, taste good, and, for certain products, have a longer shelf life. There are a number of possible scientific solutions to meet these demands, such as genetic modification and gamma irradiation, but these have met with consumer resistance. High pressure processing (HPP), also referred to as high hydrostatic pressure or ultra-high pressure processing, is one technology that has the potential to fulfill both consumer and scientific requirements. HPP uses pressures up to 900 megapascals (MPa, approximately equal to 9,000 atmospheres or 135,000 pounds per square inch) to kill many of the microorganisms found in foods, even at room temperature, without degrading vitamins and flavor and color molecules in the process (Polydera et al., 2005).

The use of HPP as a twenty-first-century food processing technology has actually evolved over many centuries, largely from many years of military research aimed at improving the effectiveness of guns and cannons. Crossland (1995) describes how these weapons developed so that they could withstand the high pressures generated from gunpowder discharges. The outcome of this work, dedicated to improving armaments, led to more widespread availability of high pressure equipment that could be adopted for a diverse variety of uses. For example, HPP has contributed significantly to aircraft safety and reliability by its use to mold turbine blades while eliminating minute flaws in their microstructure that could otherwise lead to cracks and catastrophic failure in highly stressed aero-engines. HPP has also been used for many years in the manufacture of ceramics and industrial diamonds.

Bert Hite (1899) at West Virginia University Agricultural Experimental Station published the first detailed report of the use of high pressure as a method of food preservation. He reported that milk "kept sweet for longer" after a pressure treatment of ~650 MPa for 10 min at room temperature. Hite et al. (1914) later reported that pressure could be used to extend the shelf life of fruits, concluding that fruits and fruit juices responded well to high pressure because the "yeasts and other organisms having most to do with decomposition are very susceptible to pressure, while other organisms not so susceptible do not long survive the acid media." Vegetables, however, he "abandoned as hopeless," due to the presence of sporeforming bacteria that survived the pressure treatment and could grow in the low-acid environment. Later, Cruess (1924) also proposed that high pressure could be used to successfully preserve fruit juices, where the low pH inhibited the growth of sporeformers. Other scientists studied the effects of pressure on the physical properties of foods. Bridgeman (1914) reported that high pressure coagulated egg albumen, but in a different way from the effects of heat. Payens and Heremans (1969) described the effects of pressure on β-casein molecules in milk, and Macfarlane (1973) reported that, under certain conditions, high pressure could be used to tenderize meat.

High Pressure Processing Equipment

A typical modern high pressure system consists of a pressure vessel and a pressure-generating device. Food packages are loaded into the vessel and the top is closed. The pressure medium, usually water, is pumped into the vessel from the bottom. Once the desired pressure is reached, the pumping is stopped, valves are closed, and the pressure can be maintained without further need for energy input. A fundamental principle underlying HPP is that the high pressure is applied in an isostatic manner such that all regions of the food experience a uniform pressure, unlike heat processing, where temperature gradients are established. As an isostatic process, the pressure is transmitted rapidly and uniformly throughout both the pressure medium and the food. The work of compression during HPP treatment also increases the temperature of foods through a process known as adiabatic heating, and the extent of the temperature increase varies with the composition of the food (normally 3–9°C/100 MPa). The increasing interest in producing shelf-stable pressure-treated foods requires using HPP in conjunction with heat (initial temperatures around 80–90°C, and compression heating temperatures >121°C) to kill resistant spores.

HPP is traditionally a batch process and pressure vessels used for commercial food production have capacities of 35–350 L. In the case of liquids, such as fruit juices, the vessel is filled with the juice, which acts as the pressure transmission fluid. After treatment, the juice can be transferred to an aseptic filling line, similar to that used for UHT liquids. A series of these vessels can work in a staggered sequence (one vessel filling with juice, one vessel pressurizing, and another vessel emptying) for an overall system that is semi-continuous.

Substrate Characteristics

The primary aim of treating foods with HPP in most cases is to reduce or eliminate the relevant foodborne microorganisms that may be present. The molecular composition of the food substrate, however, can significantly affect the extent to which HPP kills the inhabitant microorganisms. The application of HPP may inactivate enzymes (Butz et al., 2002) or alter the physical properties of the food material (e.g., denature structural proteins or densify texture). It is important to consider both the effects of the food substrate on slowing the microbial inactivation kinetics and the effect of the HPP on the properties of the foodstuff, when optimizing processing conditions for specific foods.

Patterson et al. (1995) reported that treating *E. coli* O157:H7 under the same conditions of 700 MPa for 30 min at 20°C resulted in a 6 log reduction in phosphate-buffered saline, a 4 log reduction in poultry meat, and a <2 log reduction in UHT milk. Other researchers have also found that microorganisms appear less sensitive to the lethal effects of high pressure when treated in certain foods, such as milk (Hauben et al., 1998). The reasons for these effects are not clear, but it may be that certain food constituents such as proteins and carbohydrates (Simpson and Gilmour, 1997) or cations such as Ca^{2+} contribute to this effect.

The pH and water activity (a_w) of foods can also significantly affect the inactivation of microorganisms by HPP. Most microorganisms tend to be more susceptible to pressure in lower pH environments, and pressure-damaged cells are less likely to survive in acidic environments. This aspect can be of commercial value for the HPP treatment of fruit juices, in which pathogens such as *E. coli* O157:H7 may survive the initial pressure treatment but die within a relatively short time during cold storage in the high-acid conditions (Linton et al., 1999).

In recent years, food preservation strategies have been developed that combine HPP with the use of anti-microbial food additives. HPP has been used to sensitize *Salmonella* (Gram-negative) or *L. monocytogenes* (Gram-positive) to nisin and lysozyme, respectively (Masschalk et al., 2001; Kalchayanand et al., 1998). Other workers (López-Pedemonte et al., 2003) formulated cheese to contain 1.56 mgL^{-1} of nisin to augment the capacity of HPP to inactivate *B. cereus* spores in the cheese. A two-step HPP protocol was used that (i) first germinated the dormant spores at 60 MPa, 30°C, 210 min, and then (ii) killed the vegetative cells at 400 MPa, 30°C, 15 min, and resulted in approximately a 2.4 $\log_{10}$ inactivation. Other potential combinations of high pressure with anti-microbial agents include the use of lacticin 3147 (Ross et al., 2000), lactoperoxidase (Garcia-Graells et al., 2003), and carvacrol (Karatzas et al., 2001).

Microbiological Aspects of HPP

As a food preservation method, the effectiveness of HPP in destroying foodborne microorganisms depends on a number of intrinsic and extrinsic factors that must be taken into account when optimizing pressure treatments for particular foods (Table 1.1).

HPP is similar to thermal processing in that there is a threshold value (specific to each microorganism) below which no inactivation occurs. Above the threshold, the lethal effect of the process tends to increase as the pressure and/or temperature increases. In instances where either increasing the magnitude or the duration of the applied pressure is unfavorable or infeasible, increasing the temperature of the HPP process can have a significant effect on the extent of microbial inactivation. Patterson and Kilpatrick (1998) found that pressures up to 700 MPa combined with mild heating up to 60°C (initial temperature) were more lethal than either treatment alone for inactivating pathogens such as *E. coli* O157:H7 and *S. aureus* in milk and poultry meat. As mentioned

Table 1.1. Response of microorganisms to high hydrostatic pressure

Microorganism	Substrate	Treatment Conditions P (MPa)/T (°C)/t (min)	Log_{10} Reduction	Source
Aeromonas hydrophila	Ground pork	253 / 25 / 15	7	Ellenberg and Hoover (1999)
Bacillus cereus	Skimmed milk	400 / 30 / 18	2.9–3.4	McClements et al. (2001)
Bacillus cereus spores	Skimmed milk	400 / 30 / 18	None	McClements et al. (2001)
Campylobacter jejuni	Pork slurry	300 / 25 / 10	6	Shigehisa et al. (1991)
C. sporogenes spores	Chicken breast	680 / 80 / 20	5	Crawford et al. (1996)
Escherichia coli (ETEC)	Skimmed milk	600 / 20 / 15	3.44	Linton et al. (2001)
		700 / 20 / 15	> 7	
E. coli O157:H7	Skimmed milk	600 / 20 / 15	4.2–6.7	Linton et al. (2001)
		700 / 20 / 15	> 7	
E. coli O157:H7	Orange juice (pH 3.4–3.9)	550 / 20 / 5	> 7	Linton et al. (1999)
E. coli O157:H7	Phosphate buffered saline (PBS)	500 / unknown / 30	< 2	Benito et al. (1999)
Hamster-adapted scrapie agent	Hamster brain homogenate	700–1,000 / 60 / 120	Increase in survival rate of hamsters	García et al. (2004)
Hepatitis A	Tissue culture medium	450 / 21 / 5	> 6 log_{10} in pfu	Kingsley et al. (2002)
	Seawater	450 / 21 / 5	< 2 log_{10} inactivation	
	PBS		< 3	

(*cont.*)

Table 1.1. *(cont.)*

Microorganism	Substrate	Treatment Conditions P (MPa)/T (°C)/t (min)	Log_{10} Reduction	Source
Listeria monocytogenes	Raw poultry meat Skimmed milk	375 / 20 / 30	6 < 2	Patterson et al. (1995)
Listeria monocytogenes	Skimmed milk	400 / 30 / 24	4	McClements et al. (2001)
Listeria monocytogenes	Citrate buffer	375 / 20 / 10	7.9	Ritz et al. (2002)
Pseudomonas fluorescens	Skimmed milk	250 / 30 / 18	5	McClements et al. (2001)
Saccharomyces cerevisiae	Pork slurry	300 / 20 / 10	2	Shigehisa et al. (1991)
Salmonella Enteritidis	PBS	500 / 20 / 15	6	Patterson et al. (1995)
Salmonella Enteritidis	Orange juice Sour cherry juice	350 / 30 / 5	> 8 > 8	Bayindirli et al. (2005)
Salmonella Senftenberg	PBS	340 / 23 / 10	4	Metrick et al. (1989)
Salmonella Typhimurium	PBS	400 / 20 / 15	> 6	Patterson et al. (1995)
Staphylococcus aureus	Raw poultry meat PBS	600 / 20 / 30	< 4 < 4	Patterson et al. (1995)
Vibrio parahaemolyticus	Oyster homogenate	300 / 10 / 2	> 6	Cook (2003)
Yersinia enterocolitica	Ground pork	304 / 25 / 15	7	Ellenberg and Hoover (1999)

earlier, high initial temperatures (>70°C) are required to kill bacterial spores and produce commercially sterile, shelf-stable foods.

Inactivation curves of HPP-treated microorganisms have been reported to exhibit very definite "tails" by various researchers (Shigehisa et al., 1991; Styles et al., 1991; Ludwig et al., 1992; Patterson et al., 1995). Metrick et al. (1989) reported tailing effects for *S.* Typhimurium and *S.* Senftenberg. When the resistant tail populations were isolated, grown, and again exposed to pressure, there was no significant difference in the pressure resistance between them and the original cultures, suggesting the tails were not simply a resistant population. The tailing phenomenon may be independent of the mechanisms of inactivation and due to clustering of groups of cells or population heterogeneity arising from genetic variation. Alternatively, tailing may be a normal feature of the mechanism of resistance involving adaptation and recovery (Earnshaw, 1995). In practice, the non-logarithmic inactivation curves make it difficult to determine the appropriate kinetics parameters.

The response of different types of microorganisms to pressure varies significantly in the following order of increasing resistance: vegetative bacteria < yeasts and molds < viruses < bacterial spores. Bacterial spores can be extremely resistant to high pressure, just as they are resistant to other lethal treatments such as heat, irradiation, and chemical agents (Gould and Sale, 1970). The notorious foodborne hazard *Clostridium botulinum* spores, especially non-proteolytic type B, are among the most pressure-resistant spores (Reddy et al., 2001).

Yeasts are generally relatively sensitive to pressure, and HPP has been used successfully to extend the shelf life of acidic products whose spoilage microflora are primarily yeasts (fruit sauces, juices, and purees). There is relatively little information on the pressure sensitivity of molds, but it has been shown that vegetative forms are relatively sensitive, while ascospores are more resistant (Voldrich et al., 2004). Butz et al., 1996, showed that the vegetative forms of the molds *Byssochlamys nivea, B. fulva, Eupenicillium* sp., and *Paecilomyces* sp. were inactivated within a few minutes using 300 MPa and 25°C. Separate HPP treatments of 800 MPa, 70°C, 10 min and 600 MPa, 10°C, 10 min were required to reduce to undetectable levels ascospore inocula (< 10^6/mL) of *B. nivea* and (10^7/mL) of *Eupenicillium*, respectively.

Giddings et al. (1929) published one of the first reports on the inactivation of viruses by high pressures using 920 MPa to inactivate

Tobacco Mosaic Virus (TMV). Otake et al. (1997) reported that pressures of 400–600 MPa for 10 min could reduce the numbers of viable particles of Human Immunodeficiency Virus (HIV) by 10^4 to 10^5. Polio virus in tissue culture medium is relatively pressure resistant, with 450 MPa at 21°C for 5 min giving no reduction in plaque forming units (PFU), whereas the same treatment conditions effected a reduction of a 6 $\log_{10}$PFU/mL stock culture of Hepatitis A to undetectable levels (Kingsley et al., 2002). Changing the medium to seawater increased the pressure resistance of the Hepatitis A virus. Calci et al. (2005) reported that HPP treatments were less effective in inactivating Hepatitis A virus in oysters compared to previously published data in culture medium. Feline Calicivirus, a Norwalk virus surrogate (Kingsley et al., 2002), and human rotavirus (Khadre and Yousef, 2002) were correspondingly more sensitive to pressure than Hepatitis A in tissue culture medium. An HPP treatment of 250 MPa at −15°C for 60 min of Foot and Mouth Disease Virus (FMDV) in 1 M urea destroyed the infectivity of FMDV but maintained the integrity of the capsid structure (Ishimaru et al., 2004).

There is evidence that high pressure may reduce the infectivity of prions in inoculated hot dogs, and thereby provide a manner in which to ensure the safety of processed meat products from bovine spongiform encephalopathy (Brown et al., 2003). Similarly, recent experiments involving infecting hamsters intracerebrally using hamster-adapted scrapie strain 263k (García et al., 2004) showed that pre-treating the prions with HPP (> 700 MPa, 60°C, 2 hr) significantly increased the survival rate of the animals.

Vegetative bacterial cells tend to be most sensitive to pressure when treated in the exponential phase of growth and most resistant in the stationary phase of growth (Mackey et al., 1995). When bacteria enter the stationary phase they can synthesize new proteins that protect the cells against a variety of adverse conditions such as high temperature, high salt concentrations, and oxidative stress. It is not known if these proteins also protect the bacteria against HPP. In some cases, a positive correlation between pressure resistance and heat resistance has been reported. For example, an *E. coli* O157:H7 strain isolated from a major hamburger patty outbreak in the United States showed less than a 1 $\log_{10}$ reduction when treated in laboratory medium at 500 MPa at < 45°C for 30 min. This strain was also more resistant to heat, acid, oxidative, and osmotic stresses than a pressure-sensitive strain (Benito et al., 1999).

However, other studies on *Salmonella* spp. report there is only a weak correlation, or no correlation, between pressure resistance and resistance to other physical treatments such as heat and irradiation (Sherry et al., 2004).

There can be significant variation in the pressure resistance between different strains of the same species, including certain strains of *L. monocytogenes* (Simpson and Gilmour 1997) and *E. coli* O157:H7 (Patterson et al., 1995). In general, the lethal effects of HPP include a number of different processes, particularly damage to the cell membrane and inactivation of key enzymes that are involved in DNA replication and transcription (for reviews see Hoover et al., 1989; Smelt et al., 2001).

Commercial HPP Foods

HPP-treated products continue to proliferate in the global marketplace largely depending on the initiatives of food companies and the positive responses of consumers to these products. Commercial food applications of HPP have focused primarily on the ability of pressure to kill spoilage organisms and relevant foodborne pathogens and extend product shelf life. While HPP successfully kills microbes and extends product shelf life, consumer preferences ultimately determine the success of individual products in the marketplace. HPP has been used for a number of successful commercial products, primarily because HPP-treated foods, in addition to being microbiologically safe, retain more of their original fresh taste, texture, and nutritional content such that these products are often superior in quality compared to their thermal processing counterparts. HPP can also confer unique benefits to products that impart them with advantages in the marketplace.

Guacamole typically has a relatively short shelf life caused by enzymatic reactions and the growth of microorganisms. Heat treatments eliminate the spoilage organisms but also induce the loss of the fresh green color that appeals to consumers. HPP (around 600 MPa for a few minutes) kills the spoilage organisms and concomitantly inactivates the relevant enzymes to yield a product featuring the sensory characteristics of the fresh product (particularly the green color) and a chilled shelf life exceeding 30 days.

HPP (treatments in the range of approximately 500–600 MPa for a few minutes) has been adopted for the post-packaging pasteurization of sliced meat products, including fermented meats, whole and sliced cured deli ham, dry cured ham, pre-cooked chicken strips, cold cuts, and other delicatessen meat products. In these instances, the HPP treatment eliminates spoilage organisms and gives additional food safety assurance by inactivating bacterial pathogens such as *L. monocytogenes*, *E. coli*, and *Salmonella* spp. The meat products retain their sensory characteristics and have a dramatic extension of shelf life under chilled storage conditions (up to 100 days). Accordingly, the food makers can reduce the levels of added chemical preservatives (e.g., sodium lactate, potassium lactate, and sodium diacetate) that create a metallic or salty aftertaste, and consequently the meat tastes fresher and better (Downie, 2005; Adamy, 2005).

HPP is successful in killing *Vibrio parahaemolyticus* and *V. vulnificus* on raw oysters. The resultant product retains the sensory characteristics of fresh oysters for an extended shelf-life period. HPP also loosens the oyster's adductor muscle and induces the shell to open of its own accord. This "self-shucking" aspect significantly reduces the need for manual shucking by hand and significantly increases the quantity of meat removed from the shell. A heat shrink plastic band is placed around each oyster shell prior to the HPP treatment to keep the shell closed during distribution and storage. HPP has also been used successfully to treat other types of shellfish, such as mussels, *Nephrops,* and lobsters, and similarly improved microbiological quality and product yields (Murchie et al., 2005).

References

Adamy, J. 2005, 17 February. High-pressure process helps keep food bacteria-free. *Wall Street Journal,* p. B1.

Bayindirli, A., H. Alpas, F. Bozoglu, and M. Hizal. 2005. Efficiency of high pressure treatment on inactivation of pathogenic microorganisms and enzymes in apple, orange, apricot and sour cherry juices. *Food Control* 17:52–58.

Benito, A., G. Ventoura, M. Casadei, T. Robinson, and B. Mackey. 1999. Variation in resistance of natural isolates of *Escherichia coli* O157 to high hydrostatic pressure, mild heat, and other stresses. *Applied and Environmental Microbiology* 65:1564–1569.

Bridgeman, P.W. 1914. The coagulation of albumen by pressure. *Journal of Biological Chemistry* 19:511–512.

Brown P., R. Meyer, F. Cardone, and M. Pocchiari. 2003. Ultra-high pressure inactivation of prion infectivity in processed meat: A practical method to prevent human infection. *Proceedings of the National Academy of Sciences* 100(10): 6093–6097.

Butz, P., R. Edenharder, A. Fernández Garçia, H. Fister, C. Merkel, and B. Tauscher. 2002. Changes in functional properties of vegetables induced by high pressure treatment. *Food Research International* 35:295–300.

Butz, P., S. Funtenberger, T. Haberditzl, and B. Tausher. 1996. High pressure inactivation of *Byssochlamys nivea* ascospores and other heat resistant moulds. *Lebensmittel-Wissenschaft und Technologie* 29:404–410.

Calci, K.R., G.K. Meade, R.C. Tezloff, and D.H. Kingsley. 2005. High-pressure inactivation of hepatitis A virus within oysters. *Applied and Environmental Microbiology* 71:339–343.

Cook, D. 2003. Sensitivity of *Vibrio* species in phosphate buffered saline and in oysters to high-pressure processing. *Journal of Food Protection* 66:2276–2282.

Crawford, Y.J., E.A. Murano, D.G. Olsen, and K. Shenoy. 1996. Use of high hydrostatic pressure and irradiation to eliminate *Clostridium sporogenes* in chicken breast. *Journal of Food Protection* 59:711–715.

Crossland, B. 1995. "The development of high pressure equipment." In: *High Pressure Processing of Foods*, ed. D.A. Ledward, D.E. Johnston, R.G. Earnshaw, and A.P.M. Hastings, pp. 7–26. Nottingham, England: Nottingham University Press.

Cruess, W.V. 1924. *Commercial Fruit and Vegetables Products*. New York: McGraw-Hill.

Downie, D. 2005, 6 June. Hormel removes preservatives. Food Safety Network, http://foodsafetynetwork.ca/fsnet/2005/6-2005/fsnet_june_7.htm#story5.

Earnshaw, R.G. 1995. "High pressure microbial inactivation kinetics." In *High Pressure Processing of Foods,* ed. D.A. Ledward, D.E. Johnston, R.G. Earnshaw, and A.P.M. Hastings, pp. 37–46. Nottingham, England: Nottingham University Press.

Ellenberg, L., and D.G. Hoover. 1999. Injury and survival of *Aeromonas hydrophila* 7965 and *Yersinia enterocolitica* 9610 from high hydrostatic pressure. *Journal of Food Safety* 19:263–276.

García, A.F., P. Heindl, H. Voight, M. Büttner, D. Wienhold, P. Butz, J. Stärke, B. Tauscher, and E. Pfaff. 2004. Reduced proteinase K resistance and infectivity of prions after pressure treatment at 60°C. *Journal of General Virology* 85:261–264.

Garcia-Graells, C., I.V. Opstal, S.C.M. Vanmuysen, and C.W. Michiels. 2003. The lactoperoxidase system increases efficacy of high-pressure inactivation of foodborne bacteria. *International Journal of Food Microbiology* 81:211–221.

Giddings, N.J., H.A. Allard, and B.H. Hite. 1929. Inactivation of tobacco mosaic virus by high pressure. *Phytopathology* 19:749–750.

Gould, G.W., and A.J.H. Sale. 1970. Initiation of germination of bacterial spores by hydrostatic pressure. *Journal of General Microbiology* 60:335–346.

Hauben, K.J.A., K. Bernaerts, and C.W. Michiels. 1998. Protective effect of calcium on inactivation of *Escherichia coli* by high hydrostatic pressure. *Journal of Applied Microbiology* 85:678–684.

Hite, B.H. 1899. The effect of pressure on the preservation of milk. *West Virginia Agricultural Experimental Station Bulletin* 58:15–35.

Hite, B.H., N.J. Giddings, and C.E. Weakly. 1914. The effect of pressure on certain micro-organisms encountered in the preservation of fruits and vegetables. *West Virginia Agriculture Experimental Station Bulletin* 146:1–67.

Hoover, D.G., C. Metrick, A.M. Papineau, D.F. Farkas, and D. Knorr. 1989. Biological effects of high hydrostatic pressure on food microorganisms. *Food Technology* 43:99–107.

Ishimaru, D., D. Sá-Carvalho, and J.L. Silva. 2004. Pressure-inactivated FMDV: A potential vaccine. *Vaccine* 22:2334–2339.

Kalchayanand, N., A. Sikes, C.P. Dunne, and B. Ray. 1998. Factors influencing death and injury of foodborne pathogens by hydrostatic pressure-pasteurization. *Food Microbiology* 15:207–214.

Karatzas, A.K., E.P.W. Kets, E.J. Smid, and M.H.J. Bennik. 2001. The combined action of carvacrol and high hydrostatic pressure on *Listeria monocytogenes* Scott A. *Journal of Applied Microbiology* 90:463–469.

Khadre, M.A., and A.E. Yousef. 2002. Susceptibility of human rotavirus to ozone, high pressure and pulsed electric field. *Journal of Food Protection* 65:1441–1446.

Kingsley, D.H., D.G. Hoover, E. Papafragkou, and G.P. Richards. 2002. Inactivation of hepatitis A virus and a calicivirus by high hydrostatic pressure. *Journal of Food Protection* 65:1605–1609.

Linton, M., J.M.J. McClements, and M.F. Patterson. 1999. Survival of *Escherichia coli* O157:H7 during storage of pressure-treated orange juice. *Journal of Food Protection* 62:1038–1040.

———. 2001. Inactivation of pathogenic *Escherichia coli* in skimmed milk using high hydrostatic pressure. *Innovative Food Science and Emerging Technologies* 2:99–104.

López-Pedemonte, T.J., A.X. Roig-Sagués, A.J. Trujillo, M. Capellas, and B. Guamis. 2003. Inactivation of spores of *Bacillus cereus* in cheese by high hydrostatic pressure with the addition of nisin or lysozyme. *Journal of Dairy Science* 86:3075–3081.

Ludwig, H., D. Bieler, K. Hallbauer, and W. Scigalla. 1992. "Inactivation of microorganisms by hydrostatic pressure." In: *High Pressure and Biotechnology*, ed. C. Balny, R. Hayashi, K. Heremans, and P. Masson. Vol. 224, pp. 25–32. London: Colloque INSERM/J. Libby Eurotext Ltd.

Macfarlane, J.J. 1973. Pre-rigor pressurization of muscle effects on pH, shear value and taste panel assessment. *Journal of Food Science* 38:294–298.

Mackey, B.M., K. Forestiere, and N. Isaacs. 1995. Factors affecting the resistance of *Listeria monocytogenes* to high hydrostatic pressure. *Food Biotechnology* 9: 1–11.

Masschalk, B.R., R. Van Houdt, E.G.R. Van Haver, and C. Michiels. 2001. Inactivation of Gram-negative bacteria by lysozyme, denatured lysozyme and lysozyme derived peptides under high pressure. *Applied and Environmental Microbiology* 67:339–344.

McClements, J.M.J., M.F. Patterson, and M. Linton. 2001. The effect of stage of growth and growth temperature on high hydrostatic pressure inactivation of some psychotropic bacteria in milk. *Journal of Food Protection* 64:514–522.

Metrick, C., D.G. Hoover, and D.F. Farkas. 1989. Effects of high hydrostatic pressure on heat resistant and heat sensitive strains of *Salmonella*. *J. Food Science* 54:1547–1564.

Murchie, L.W., M. Cruz-Romero, J.P. Kerry, M. Linton, M.F. Patterson, M. Smiddy, and A.L. Kelly. 2005. High pressure processing of shellfish: A review of microbiological and other quality aspects. *Innovative Food Science and Emerging Technologies* 6:257–270.

Otake, T., H. Mori, T. Kawahata, Y. Izumoto, H. Nishimura, I. Oishi, T. Shigehisa, and H. Ohno. 1997. "Effects of high hydrostatic pressure treatment on HIV infectivity." In *High Pressure Research in the Biosciences*, ed. K. Heremans, pp. 233–236. Leuven, Belgium: Leuven University Press.

Patterson, M.F., and D.J. Kilpatrick. 1998. The combined effect of high hydrostatic pressure and mild heat on inactivation of pathogens in milk and poultry. *Journal of Food Protection* 61:432–436.

Patterson, M.F., M. Quinn, R. Simpson, and A. Gilmour. 1995. Sensitivity of vegetative pathogens to high hydrostatic pressure treatment in phosphate-buffered saline and foods. *Journal of Food Protection* 58:524–529.

Payens, T.A.J., and K. Heremans. 1969. Effect of pressure on the temperature-dependent association of β-casein. *Biopolymers* 8:335–345.

Polydera, A.C., N.G. Stoforos, and P.S. Taoukis. 2005. Effect of high hydrostatic pressure treatment on post processing antioxidant activity of fresh navel orange juice. *Food Chemistry* 91:495–503.

Reddy, N.R., H.M. Solomon, R.C. Telzloff, V.M. Balasubramaniam, E.J. Rhodehamel, and E.Y. Ting. 2001. Inactivation of *Clostridium botulinum* spores by high pressure processing. *2001 Annual Report of the National Centre for Food Safety and Technology*, Summit-Argo, IL. Cited by Sizer, C.E., V.M. Balasubramaniam, and E. Ting. 2002. Validating high pressure processing for low acid foods. *Food Technology* 56:36–42.

Ritz M., J.L. Tholozan, M. Federighi, and M.F. Pilet. 2002. Physiological damages of *Listeria monocytogenes* treated by high hydrostatic pressure. *International Journal of Food Microbiology* 79:47–53.

Ross, R.P, T. Beresford, C. Hill, and S.M. Morgan. 2000. Combination of hydrostatic pressure and lacticin 3146 causes increased killing of *Staphylococcus* and *Listeria*. *Journal of Applied Microbiology* 88:414–420.

Sherry, A.E., M.F. Patterson, and R.H. Madden. 2004. Comparison of 40 *Salmonella enterica* serovars injured by thermal, high pressure and irradiation stresses. *Journal of Applied Microbiology* 96:887–893.

Shigehisa, T., T. Ohmori, A. Saito, S. Taji, and R. Hayashi. 1991. Effects of high pressure on the characteristics of pork slurries and the inactivation of micro-organisms associated with meat and meat products. *International Journal of Food Microbiology* 12:207–216.

Simpson, R.K., and A. Gilmour. 1997. The effect of high hydrostatic pressure on the activity of intracellular enzymes of *Listeria monocytogenes*. *Letters in Applied Microbiology* 25:48–53.

Smelt, J.P.M., J.C. Hellemons, and M.F. Patterson. 2001. "Effects of high pressure on vegetative microorganisms." In: *Ultra High Pressure Treatments of Foods*, ed. M.E.G. Hendrickx and D. Knorr, pp. 55–76. New York: Kluwer Academic/Plenum Publishers.

Styles, M.F., D.G. Hoover, and D.F. Farkas. 1991. Response of *Listeria monocytogenes* and *Vibrio parahaemolyticus* to high hydrostatic pressure. *Journal of Food Science* 56:1404–1407.

Voldrich, M., J. Dobias, L. Ticha, M. Cerovsky, and J. Kratka. 2004. Resistance of vegetative cells and ascospores of heat resistant mould *Talaromyces avellaneus* to the high pressure treatment in apple juice. *Journal of Food Engineering* 61: 541–543.

Chapter 2

Germination of Spores of *Bacillus subtilis* by High Pressure

Peter Setlow

Introduction

Inactivation of bacterial spores is one of the major challenges in food sterilization and preservation. This is especially so in the case of minimally processed ready-to-eat foods. While spores of all *Bacillus* and *Clostridium* species can be readily destroyed by high temperature treatments, such treatments can have detrimental effects on food quality. Consequently, when high pressure (HP) was shown to inactivate spores, this raised the possibility that such a treatment might be a useful alternative to high temperature processing. HP has been especially attractive in this regard, since it has many less deleterious effects than high temperatures on food quality.

Early work on inactivation of spores by HP established that with pressures of 50–900 megaPascals (MPa), spores are first germinated and then inactivated, either by the pressure treatment, or, more often, by the temperature of the HP treatment (Gould and Sale, 1969; Sale et al., 1970; Knorr, 1999; Raso and Barbosa-Canovas, 2003). While the germinated spores may be killed by high temperature treatment, this method of inactivation requires much lower temperatures than the killing of dormant spores (Nicholson et al., 2000), with a resultant improvement in food qualities. Consequently, HP has significant potential as a value-added food processing procedure.

Spore inactivation as a result of HP treatment involves the triggering of at least some spore germination events. Thus it is appropriate to understand how HP triggers spore germination, with an eye toward making this process more efficient. The understanding of this process requires an understanding of spore physiology and structure as well as knowledge of the mechanisms of spore germination by stimuli other than HP, since HP likely co-opts some physiological germination pathway(s) to trigger spore germination. Consequently, I will initially discuss spore properties and structure, then what is known about spore germination in response to stimuli other than HP, and then what is known about spore germination by HP. I will end with a listing of some unanswered questions about HP germination.

Spore Properties and Structure

General Properties

Spores of *Bacillus* and *Clostridium* species are formed in sporulation, a process triggered by starvation and perhaps environmental stress (Driks, 2002a). The spores are released into the environment upon lysis of the sporangium and they are metabolically dormant, exhibiting no detectable metabolism of endogenous or exogenous compounds. Dormant spores lack common high-energy compounds such as ATP, although the "low-energy" forms of such molecules (e.g., AMP) are present (Setlow, 1994; and see below). In addition to their dormancy, spores are also extremely resistant to a wide variety of environmental stresses including wet heat, desiccation, dry heat, freezing and thawing, UV radiation, γ-radiation, and many toxic chemicals including acids, bases, aldehydes, oxidizing agents, alkylating agents, and aliphatic and aromatic alcohols (Nicholson et al., 2000; Setlow, 2006). Dormant spore resistance is much greater than the resistance of the germinated spore or growing cell of the same strain. For example, dormant spores of *Bacillus* species are generally resistant to temperatures 40–45°C higher than are growing cells (Warth, 1978, 1980). Dormant spore resistance to other treatments is equally impressive, as spores are not killed by repeated cycles of desiccation and rehydration, while growing cells rapidly lose viability upon desiccation unless specially protected (Nicholson et al., 2000; Setlow, 2006).

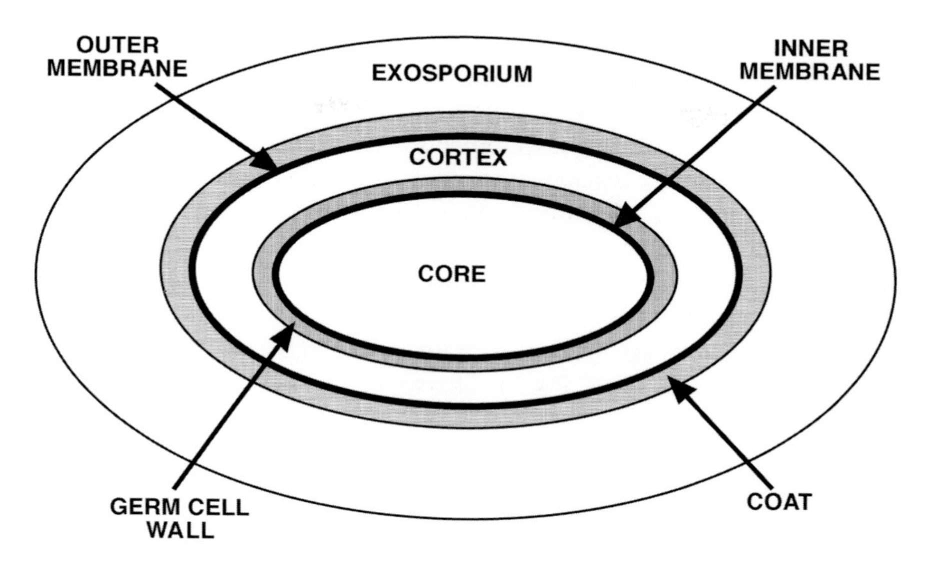

Figure 2.1. Dormant spore structure. The various layers of the dormant spore are not drawn to scale.

Spore Structure

Spore resistance to the many treatments noted above is due to a number of features unique to the dormant spore, including the spore's unique structure as well as proteins and small molecules unique to spores. Starting from the outside, the spore structure consists of a number of layers including exosporium, coat, outer membrane, cortex, germ cell wall, inner membrane, and central core (Figure 2.1).

Exosporium

The exosporium is a large balloon-like layer present in spores of many species, in particular those of the *B. cereus* group. In contrast, spores of a number of species, including *B. subtilis*, do not have an exosporium, or it is only a minor part of the outer layer of the spore coat (Driks, 1999). The exosporium contains carbohydrate and protein, including some proteins unique to spores (Driks, 1999; Redmond et al., 2004). However, the exosporium likely has no significant role in spore resistance or spore germination.

Coat

The spore coat is made up largely of protein, with ≥ 50 proteins unique to the *B. subtilis* spore coat (Driks, 1999, 2000b; Kim et al., 2006). The coat blocks access of lytic enzymes to peptidoglycan (PG) layers below and is important in spore resistance to many chemicals, perhaps by serving as a "reactive armor" to detoxify such chemicals. However, the coats are not important in HP resistance (Paidhungat et al., 2002). Spore proteins located in or adjacent to the spore coat play roles in spore germination (Driks, 1999, 2000b; Bagyan and Setlow, 2002; Chirakkal et al., 2002). However, the coats are probably not important in germination triggered by HP (Paidhungat et al., 2002). The spore coats are normally shed at some point in spore outgrowth.

Outer Membrane

The next layer, the outer spore membrane, is extremely important in spore formation, as the engulfment of the developing spore by the mother cell encloses the developing spore in two membranes, the outer and inner membrane. Each of these membranes plays important roles in spore formation. However, the outer membrane may not be a functional membrane in dormant spores, and its removal has no effect on spore resistance to or germination by HP (Paidhungat et al., 2002).

Cortex

Lying under the outer membrane is the spore cortex. This large layer is composed of PG with a structure similar to that of vegetative cell wall PG, but with several spore-specific modifications (Popham, 2002). In the dormant spore the cortex appears to act as a straitjacket, restricting the expansion of the spore core. The cortex may also be involved in reducing the amount of water present in the spore core during spore formation (see below). The cortex is degraded in the first minutes of spore germination, and degradation of the cortex is essential to allow a germinated spore to initiate metabolism and grow (see below).

Germ Cell Wall

The next layer, the germ cell wall, is also composed of PG. However, the germ cell wall PG appears to have a structure identical to that of vegetative cell wall PG (Popham, 2002). The germ cell wall is not degraded in germination, and becomes the cell wall of the outgrowing spore (Setlow, 2003).

Inner Membrane

The spore's inner membrane has a number of unique properties, even though its phospholipid and fatty acid composition is not dramatically different from that of the corresponding growing cell (Bertsch et al., 1969; Cortezzo et al., 2004b,; Cortezzo and Setlow, 2005). The unique properties of the inner membrane include the following: (i) Molecules in the inner membrane appear to be largely immobile as determined by analysis using fluorescence redistribution after photobleaching (FRAP) of a lipid probe in this membrane (Cowan et al., 2004). This immobility of molecules in the inner membrane is lost when spores complete spore germination. (ii) The inner membrane appears to be significantly compressed in some fashion, as the volume encompassed by this membrane increases $\geq$ 2-fold upon completion of spore germination. This increase in the volume encompassed by the inner membrane takes place in the absence of synthesis of membrane components and production of ATP (Cowan et al., 2004). (iii) The passive (i.e., non-carrier mediated) permeability of the inner membrane is extremely low. Even molecules as small as methylamine, and probably even water, pass through this membrane extremely slowly (Westphal et al., 2003; Cortezzo and Setlow, 2005). The low permeability of this membrane is a major factor in dormant spore resistance to lethal agents that must cross this membrane to damage the DNA in the spore core (Cortezzo and Setlow, 2005). Another reason for the importance of the unique properties of the spore's inner membrane is that proteins that recognize nutrients that trigger spore germination are located here, as are proteins that likely play roles in ion movement early in spore germination.

Core

The central spore layer is the core. The core contains the spore's DNA, RNA, and most enzymes and is analogous to a cell's protoplast. A major feature of the spore core is its low water content. Water comprises ~80% of the wet weight of a growing cell's protoplast, and the spore's exosporium, coat, cortex, and germ cell wall also have this high level of water (Gerhardt and Marquis, 1989). However, the cores of spores of various *Bacillus* species have only 25–55% of their wet weight as water (Gerhardt and Marquis, 1989). This low core water content is attained late in spore formation in a process that involves action of the spore cortex, but how the low spore core water content is achieved is not known. The low core water content is a major factor in determining a spore's resistance to wet heat in general, as the lower the core water content the higher the spore resistance to wet heat (Gerhardt and Marquis, 1989). The core water content appears to be so low that protein mobility in the core is greatly restricted or absent (Cowan et al., 2003). Consequently, the core's low water content is likely the major reason for the spore's metabolic dormancy, since water levels in the dormant spore core appear to be too low for enzyme action. However, the amount of core water that is free (versus bound) water is not known. Although the dormant spore's core water content is low, its water content rises during germination. Upon completion of germination (see below), the percentage of core wet weight as water rises to ~80%, a value similar to that in growing cells (Gerhardt and Marquis, 1989; Popham et al., 1996).

In addition to low levels of water, the spore core has extremely low levels of common "high-energy" compounds such as ATP and other ribonucleoside triphosphates, reduced pyridine nucleotides, and acetyl-Coenzyme A (Setlow, 1994). These latter compounds disappear from the developing spore late in sporulation, but the spore does have significant levels of the "low-energy" forms of these compounds, including AMP and other ribonucleoside monophosphates, oxidized pyridine nucleotides, and Coenzyme A (Setlow, 1994). Upon completion of spore germination, when spore enzymatic activity and energy metabolism begin, the spore again accumulates high-energy compounds (Setlow, 1994).

The spore core also has high levels of several components found only in spores. One is a group of small, acid-soluble proteins (SASP) that comprise 7–15% of spore core protein. These small (55–95 aa) proteins are made in the developing spore late in sporulation and are degraded to amino acids early in spore outgrowth (Setlow, 1995; Driks, 2002b). This degradation provides amino acids for energy metabolism and protein synthesis at this period of development. There are two types of SASP in spores of *Bacillus* species, the $\alpha\beta$- and γ-type. There is a single γ-type SASP in spores, and this protein does not appear to be bound to any core component. The γ-type SASP are not found in spores of *Clostridium* species and the sequences of these 75–95 residue proteins are only moderately conserved among *Bacillus* species. The only function of γ-type SASP appears to be providing a reservoir of amino acids for use in spore outgrowth.

In contrast to the γ-type SASP, the $\alpha\beta$-type SASP, named for the two major proteins of this type in *B. subtilis* spores, is present in spores of *Clostridium* as well as *Bacillus* species. There are multiple (up to seven) $\alpha\beta$-type SASP in spores, each encoded by a unique gene. The amino acid sequences of $\alpha\beta$-type SASP are highly conserved both within and among species. However, these sequences show no notable homology to those of other proteins in current databases, and contain no motifs found in other proteins. Like γ-type SASP, the $\alpha\beta$-type SASP serve as an amino acid reservoir by their degradation early in spore outgrowth. However, the $\alpha\beta$-type SASP have an even more important function, as they bind to and saturate spore DNA. This binding protects the spore DNA from damage due to a variety of agents including wet heat, dry heat, desiccation, and many genotoxic chemicals including hydrogen peroxide, formaldehyde, and nitrous acid (Setlow, 1995, 2006; Nicholson et al., 2000). The binding of $\alpha\beta$-type SASP also alters the UV photochemistry of spore DNA (Nicholson et al., 2000; Setlow, 2001). This plays a major role in spore resistance to UV radiation. Spores lacking the majority of their $\alpha\beta$-type SASP (termed $\alpha\beta$ spores) can be readily generated, and these have exhibited decreased resistance to UV light but germinate normally, including in response to HP (Black et al., 2007). While SASP levels are high in dormant spores, these proteins are degraded early in spore outgrowth (Setlow, 1995). Their degradation is initiated by a spore-specific protease termed GPR. This protease is present in spores in a form that is active in vitro. However, GPR does not act in the dormant spore, presumably because the low core water level prevents enzyme action.

In addition to the SASP, another unique core component is pyridine-2,6-dicarboxylic acid (common name dipicolinic acid, and abbreviated DPA. See Figure 2.2 for molecular structure).

DPA makes up ~20% of core dry weight and is likely present in the core in a 1:1 complex with divalent cations, predominantly Ca^{2+}. As a consequence of the core's low core water and high solute (Ca^{2+}-DPA) content, it has been suggested that the core is in a glass-like state (Ablett et al., 1999), although this suggestion remains controversial (Leuschner and Lillford, 2003). DPA is synthesized in the mother cell late in sporulation and is taken up into the developing spore, perhaps using an uptake apparatus composed at least in part of proteins encoded by the *spoVA* operon (Errington, 1993; Tovar-Rojo et al., 2002; Vepachedu and Setlow, 2004). The spore's DPA and associated cations are excreted early

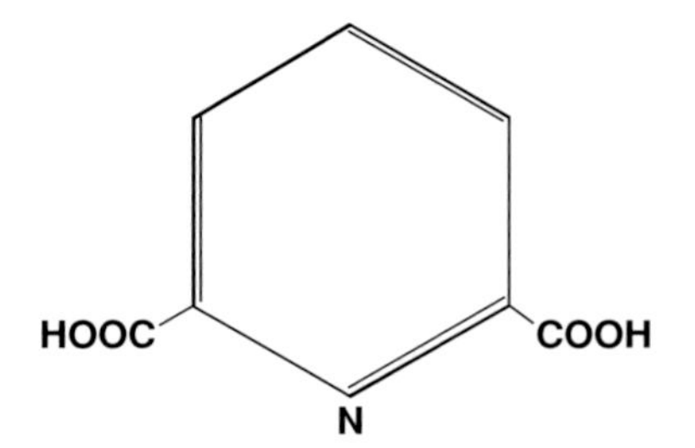

Figure 2.2. Dipicolinic acid (DPA) molecular structure.

in spore germination. While the mechanism for Ca^{2+}-DPA excretion is not known, it again may utilize some of the SpoVA proteins (Vepachedu and Setlow, 2004).

The high level of Ca^{2+}-DPA in the spore core plays a significant role in several spore resistance properties. In wet heat resistance the accumulated Ca^{2+}-DPA displaces core water, thus lowering the core water content. During dry heat, desiccation, or UV treatment, Ca^{2+}-DPA has direct protective effects on spore DNA (Paidhungat et al., 2000; Douki et al., 2005; Setlow et al., 2006). DPA is also important in stabilizing the dormant spore state in some fashion, as DPA-less spores germinate spontaneously and often lyse (Paidhungat et al., 2000, 2001). However, stable DPA-less spores can be generated by combining a mutation in the *spoVF* operon (encoding DPA synthetase) with a mutation or mutations in genes encoding components of the spore's germination apparatus (Paidhungat et al., 2000, 2001).

Spore Germination by Agents Other than HP

Spores can remain in their dormant, resistant state for many, many years, and there are reports that spores can survive for millions of years (Kennedy et al., 1994; Cano and Borucki, 1995; Vreeland et al., 2000). However, dormant spores are constantly sensing their environment. Consequently, when presented with the proper stimulus they can rapidly return to life via spore germination followed by outgrowth (Gould, 1969; Moir et al., 2002; Paidhungat and Setlow, 2002; Setlow, 2003) (Figure 2.3).

The germination process can take only a few minutes for an individual spore, but because of heterogeneity in these times between individual

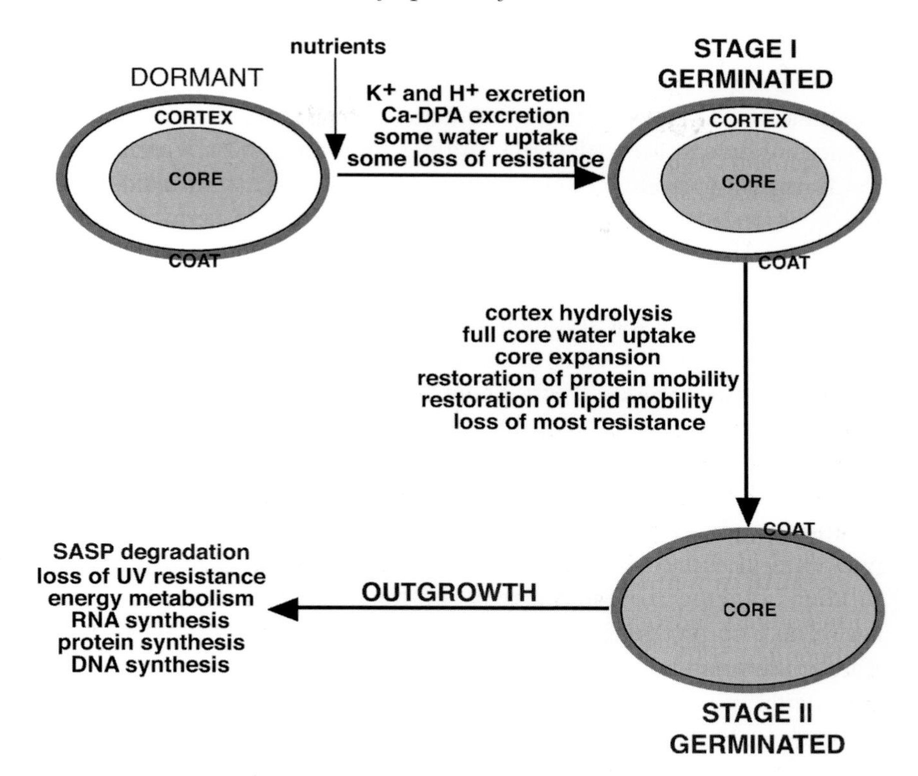

Figure 2.3. Stages in spore germination (this figure was adapted from Figure 1 of Setlow, 2003).

spores, the majority of a spore population may take much longer to complete germination. Additionally there are almost always some spores in a population, sometimes called super-dormant spores, that take days longer than the great majority of a population to complete germination (Gould, 1969). Unfortunately, the reasons for the phenomenon of spore super-dormancy are not known.

A simple distinction between the germination and outgrowth phases of a dormant spore's return to life is that germination requires neither metabolism of endogenous or exogenous compounds, direct expenditure of high-energy compounds, nor even enzyme action in the spore core (Moir et al., 2002; Paidhungat and Setlow, 2002; Setlow, 2003). In contrast, outgrowth requires enzyme action in the spore core, including energy metabolism and macromolecular synthesis. Eventually outgrowth blends into the resumption of vegetative growth. The germination process is initiated under physiological conditions by nutrients, although other types of compounds can also initiate germination, including Ca^{2+}-DPA, cationic surfactants, lysozyme or other lytic enzymes, and in spores of some strains, inorganic salts (Gould, 1969; Paidhungat and Setlow, 2002; Setlow, 2003).

Nutrients

The major physiological signal for spore germination is undoubtedly nutrients, as their presence in the environment signals that conditions are again favorable for growth. There are species- and strain-specific differences in the specific nutrients that trigger spore germination, likely reflecting differences in the metabolic capacity and ecological niches of different organisms (Gould, 1969). Common nutrient germinants are L-amino acids, D-sugars, and purine nucleosides. In many cases combinations of nutrients have synergistic effects on germination. Metabolism of nutrient germinants is not required for their triggering of germination (Moir et al., 2002; Paidhungat and Setlow, 2002; Setlow, 2003). Rather the nutrients are sensed by receptors located in the spore's inner membrane. These receptors, termed GerA homologs after the GerA receptor in *B. subtilis* spores, are encoded by tricistronic operons expressed only in the developing spore late in sporulation (Moir et al., 2002; Paidhungat and Setlow, 2002; Setlow, 2003). All three proteins of the operon are required for receptor function, and it is likely that all three proteins physically interact (Moir et al., 2002; Setlow, 2003; Igarashi and Setlow, 2005). The C protein component of most GerA type receptors undergoes covalent lipid addition on a cysteine residue near the protein's N-terminus (Moir et al., 2002). This lipid addition is essential for normal function of most GerA receptor homologs (Igarashi et al., 2004).

There are three to seven GerA type receptors in *Bacillus* and *Clostridium* spores that have been studied in sufficient detail. These different receptors are composed of three proteins, termed A, B, and C, and each type of protein shows significant sequence homology between different receptors (Moir et al., 2002; Paidhungat and Setlow, 2002; Setlow, 2003). In *B. subtilis,* where these receptors have been best studied, the GerA receptor recognizes L-alanine and L-valine, while the GerB and GerK receptors cooperate in some fashion to recognize a mixture of D-glucose, D-fructose, K^+, and one of several amino acids including

L-alanine, L-asparagine, L-serine, and L-threonine (Moir et al., 2002; Setlow, 2003; Atluri et al., 2006). While individual receptors can act alone (Moir et al., 2002; Atluri et al., 2006), proteins from different receptors can also interact (Igarashi and Setlow, 2005). However, the mechanism of receptor "co-operation" is not known. Nor is it known how binding of a nutrient germinant to a receptor triggers subsequent events in spore germination.

The initial events after binding of nutrient germinants to their receptor have been studied best in spores of *B. megaterium* and *B. subtilis* and are, in order: (i) commitment, after which nutrient germinants can be removed yet germination will proceed; (ii) release of monovalent ions such as H^+, Na^+, and K^+ and also Zn^{2+}; (iii) release of Ca^{2+}-DPA and uptake of water; (iv) hydrolysis of the cortex; and (v) further water uptake and swelling of the spore core (Gould, 1969; Moir et al., 2002; Paidhungat and Setlow, 2002; Setlow, 2003). With the last event, the core water content returns to that of a growing cell, and protein mobility and enzymatic activity in the core is restored and outgrowth begins (Paidhungat and Setlow, 2002; Cowan et al., 2003; Setlow, 2003).

The release of ions and Ca^{2+}-DPA triggered by nutrient- germinant receptor interaction is a crucial step in germination, but unfortunately it is not well understood. The composition of channels, if there are such, for release of ions and Ca^{2+}-DPA are not known with certainty, although SpoVA proteins may make up the DPA channel (Tovar-Rojo et al., 2002; Vepachedu and Setlow, 2004, 2005). The mechanism whereby such channels are "gated" is also not known. Given the unusual properties and structure of the spore's inner membrane, the environment in which channel proteins and nutrient germinant receptors must function is different from the environment in a growing cell membrane. How this difference affects protein function is, however, not known.

In the release of Ca^{2+}-DPA, $\sim$20% of the core's dry weight is lost. This loss in core dry weight is balanced by water uptake as the core water content rises significantly (Setlow et al., 2001). However, the specific mechanism(s) for water movement into the spore is not clear, and *Bacillus* species lack obvious genes encoding aquaporins. With *B. subtilis* spores this latter step, completion of which finishes stage I of germination, raises the core water content from $\sim$35% to $\sim$45% of wet weight (Setlow et al., 2001). While these stage I germinated spores remain metabolically dormant, they are much more sensitive to wet heat than are dormant spores. This characteristic may be important in the inactivation of spores by HP at elevated temperatures (see below). While stage I germinated spores normally rapidly progress further in germination, stage I germinated spores can be isolated after chemical pretreatment of dormant spores or by using spores of strains with mutations that block further progression in germination (Setlow et al., 2001).

The events in stage I of germination trigger the next steps that allow spores to complete germination by proceeding through stage II. Major events in stage II are the degradation of the PG cortex and release of cortical fragments, and the uptake of water concomitant with expansion of the germ cell wall. Spores of all species have multiple redundant enzymes involved in cortex hydrolysis (Makino and Moriyama, 2002; Setlow, 2003). Where studied, these enzymes specifically recognize a spore-specific cortical PG specific modification, termed muramic acid-δ-lactam (Popham et al., 1996). In *B. subtilis* there are two redundant cortex-lytic enzymes (CLEs), termed SleB and CwlJ, either of which is sufficient for cortex hydrolysis (Bagyan and Setlow, 2002; Chirakkal et al., 2002; Setlow, 2003). *B. subtilis* spores that lack both of their CLEs (*cwlJ sleB* mutant spores) do not give rise to colonies, as they cannot escape from the cortical straitjacket. However, their spores can be recovered by lysozyme degradation of the cortex (see below). In dormant spores both CwlJ and SleB are in a form that could be active, yet these enzymes do not work until their action is triggered by events in stage I of germination. CwlJ action is triggered by Ca^{2+}-DPA as it is released in stage I, and can also be activated by exogenous Ca^{2+}-DPA (Paidhungat et al., 2001). SleB is not activated by Ca^{2+}-DPA, but likely by some change in the physical state of cortical PG as a result of Ca^{2+}-DPA release. However, the specifics of this physical change are not known. Spores of *Clostridium* species lack a CwlJ-like enzyme but contain a SleB homolog and also have several other candidate CLEs, at least one of which is activated by proteolysis in the first minutes of germination (Makino and Moriyama, 2002). However, the role of these enzymes in germination of *Clostridium* spores has not been established.

Upon cortex degradation, the germ cell wall expands in some fashion, and this event coupled with water uptake increases the core volume ~2-fold and raises the core water content to that of a growing cell (Setlow, 2003). This allows resumption of protein mobility in the core, lipid mobility in the inner membrane, and initiation of SASP degradation and metabolism followed by macromolecular synthesis (Setlow, 2003; Cowan et al., 2003, 2004). These latter enzymatically driven events are termed outgrowth, and with completion of DNA replication the outgrowing spore is once again a growing cell.

Ca^{2+}-DPA

Ca^{2+}-DPA is a near universal spore germinant, as 20–40 mM Ca^{2+}-DPA germinates spores of most species in 30–60 min, although these high concentrations of Ca^{2+}-DPA are required for rapid germination. With spores of *B. subtilis,* exogenous Ca^{2+}-DPA activates CwlJ, which initiates cortex hydrolysis leading to spore germination. Ion and endogenous Ca^{2+}-DPA release are not triggered by nutrient binding to GerA receptor homologs in this situation, but must still leave the spore during or after cortex hydrolysis. However, the mechanism and pathway for the release of these core constituents during Ca^{2+}-DPA germination are not known.

The facile germination of spores by exogenous Ca^{2+}-DPA raises the possibility that the Ca^{2+}-DPA released upon germination of some spores in a population may trigger the germination of other spores in the population. While this could occur in very concentrated spore suspensions, the concentrations of exogenous Ca^{2+}-DPA needed to trigger spore germination efficiently are very high (10s of mM). Consequently, exogenous Ca^{2+}-DPA is an unlikely germinant in natural situations.

Cationic Surfactants

Over 40 years ago a large number of cationic surfactants were shown to trigger germination of spores of both *Bacillus* and *Clostridium* species (Rode and Foster, 1960a, 1961). One of the most effective of these agents was dodecylamine (Rode and Foster 1961). More recent work has shown that dodecylamine does not act through the GerA receptor homologs and does not activate CLEs (Setlow et al., 2003). It appears likely that this surfactant causes germination by triggering Ca^{2+}-DPA release. This could be by either activating a normally existing Ca^{2+}-DPA channel in the spore's inner membrane, or by acting on the inner membrane to directly create a channel or pore for Ca^{2+}-DPA.

Other Germinants

In addition to the germinants noted above, there are a few other triggers of spore germination. One that is likely universal is the hydrolysis of the cortex by an exogenous lytic enzyme such as lysozyme. This is usually ineffective with intact spores, as the coat normally prevents access of exogenous enzymes to the cortex. However, chemically decoated or genetically coatless spores can be readily germinated with lysozyme. If this lysozyme treatment is carried out in a hypertonic medium, viable spores are recovered in high yield (Popham et al., 1996; Setlow et al., 2001). Again, how ions and Ca^{2+}-DPA leave the spore core upon lysozyme germination is not clear.

Another procedure that can trigger spore germination is mechanical abrasion (Rode and Foster, 1960b; Jones et al., 2005). With *B. subtilis* spores, abrasion appears to trigger germination by activating either CwlJ or SleB, perhaps by some physical damage to cortical PG (Jones et al., 2005).

While the germinants noted above work well with spores of almost all species, salts alone can germinate spores of a few strains. The most well-studied spores that germinate with salts (KBr is an effective one) are those of some strains of *B. megaterium* (Gould, 1969; Paidhungat and Setlow, 2002). Various salts or ions are also co-germinants with various nutrient germinants for spores of several species (Gould, 1969; Moir et al., 2002). While the mechanism of spore germination by salt alone is not clear, this may involve the nutrient germinant receptors in some fashion (Cortezzo et al., 2004a).

Spore Germination by HP

In addition to responding to the germinants described in the previous section, spores of *Bacillus* and *Clostridium* species are also germinated by HP. The pressure levels that trigger spore germination are extremely high, ranging between 50 and 800 MPa. However, spore inactivation by HP alone is generally inefficient and requires elevated temperatures ($T > 100°C$) coupled with HP. The mechanism whereby HP triggers spore germination differs considerably at lower (50–300 MPa) and higher (300–800 MPa) pressures. Lower pressures (LP) trigger spore germination by activating the nutrient germinant receptors, while even higher pressures (EHP) cause DPA release, which then triggers subsequent events in germination (Wuytack et al., 1998, 2000; Paidhungat et al., 2002). As is not surprising, there is some overlap in the pressures that trigger germination by these two mechanisms, even though one or the other will predominate at any given pressure. However, each of these mechanisms will be discussed separately. Most detailed mechanistic information on spore germination by HP has been obtained with *B. subtilis* spores, because of the wealth of genetic and molecular genetic information and resources available with this organism.

LP Germination

Evidence that LP triggers spore germination through nutrient germinant receptors comes from the use of *B. subtilis* spores lacking several or all of these receptors (Wuytack et al., 2000; Paidhungat et al., 2002). These receptor-less spores are germinated extremely poorly by a pressure of 150 MPa in comparison to the LP germination of wild-type spores (Black et al., 2005) (Figure 2.4).

Mutations that block the lipid addition to the C-proteins of the spore's nutrient germinant receptors that is essential for receptor function also block spore germination by a pressure of 150 MPa (Black et al., 2005). Different nutrient germinant receptors also exhibit very different responsiveness to LP, and elevation of individual nutrient receptor levels increases spore germination with LP (Black et al., 2005).

Spore germination triggered by LP appears to proceed via the same pathway as nutrient germination, including Ca^{2+}-DPA release followed by cortex hydrolysis (Wuytack et al., 1998; Paidhungat et al., 2002). With *B. subtilis* spores either CwlJ or SleB are needed for cortex lysis triggered by LP treatment, but spores lacking both CwlJ and SleB release their DPA upon LP treatment (Paidhungat et al., 2002). Following cortex lysis, the LP germinated spore goes through at least the early stages of outgrowth, including SASP degradation and ATP synthesis (Wuytack et al., 1998). Spore germination by LP proceeds through stage II of germination at moderate temperatures (20–45°C). However, it appears likely that spores treated with LP at much higher temperatures (70–80°C) do not initiate let alone complete germination efficiently, perhaps because of the temperature sensitivity of nutrient germinant receptors and CLEs (Black et al., 2007).

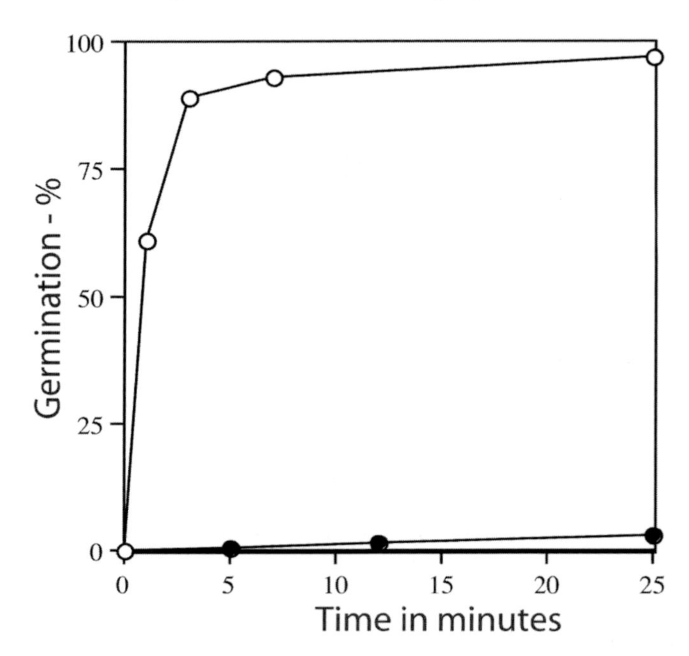

Figure 2.4. LP germination of *B. subtilis* spores with or without nutrient germinant receptors. Spores of the isogenic *B. subtilis* strains PS533 (has all nutrient germinant receptors) or FB72 (lacks all functional nutrient germinant receptors; see Paidhungat and Setlow, 2000) at an optical density of 1 were treated with a pressure of 150 MPa at 37°C in 50 mM Tris-HCl (pH 7.5). The percentage of spore germination was determined by flow cytometry after staining the spores with the nucleic acid stain Syto 16 (Molecular Probes, Eugene, OR). The data for this figure are from Black et al., 2005. The symbols used are open circles (o) to designate the response of PS533 spores, and filled circles (•) to represent FB72 spores.

Since LP triggers spore germination by activating the nutrient germinant receptors in the spore's inner membrane, the properties of this membrane may influence spore germination by LP. The spore's inner membrane has a number of unusual properties as noted above, including low permeability and lipid immobility. Thus this membrane likely has very low fluidity. Membrane fluidity in growing cells generally changes in response to growth temperatures, as membranes become more fluid at lower growth temperatures through changes in the fatty acid composition of membrane phospholipids. Spore inner membrane fatty acid composition does change as a function of growth temperature (Cortezzo et al., 2004b, 2005), as does the responsiveness of spores to LP (Raso

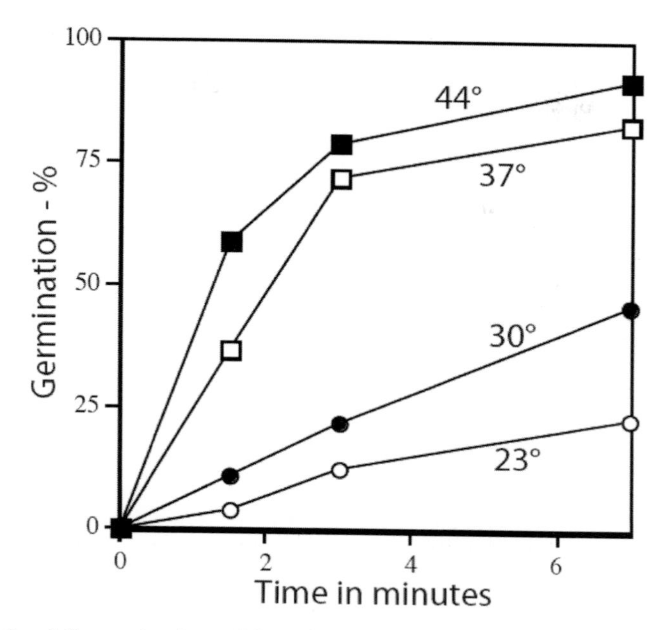

Figure 2.5. LP germination of *B. subtilis* spores made at different temperatures. Spores of *B. subtilis* strain PS533 (wild-type) were prepared at different temperatures (Melly et al., 2002), treated with a pressure of 150 MPa, and the percentage of spore germination assessed as described in the legend to figure 2.4. The data are from Black et al., 2005. The symbols used are filled squares (■) for spores made at 44°C; open squares (□) for spores made at 37°C; filled circles (•) for spores made at 30°C; and open circles (o) for spores made at 23°C.

et al., 1998a, 1998b; Black et al., 2005). *B. subtilis* spores formed at low temperatures, where membrane fluidity should be higher, germinate slower with LP than do spores produced at higher temperatures (Raso et al., 1998a, 1998b; Black et al., 2005) (Figure 2.5).

While it is tempting to infer a causal connection between likely changes in spore inner membrane fluidity as a function of temperature and changes in the rate of spore germination by LP, this connection has not been proven. However, the level of unsaturated fatty acids in the spore's inner membrane is not important in determining rates of LP spore germination, nor is oxidative damage of some kind to the spore's inner membrane (Cortezzo et al., 2004b; Black et al., 2005).

As noted above, spore germination with LP proceeds via activation of nutrient germinant receptors, and then follows the normal germination pathway outlined in the section "Spore Germination by Agents Other than HP." Thus, GerD, required for nutrient germination by *B. subtilis* spores, is also required for LP germination (Wuytack et al., 2000). Spore Ca^{2+}-DPA is also required for LP germination of *B. subtilis* spores that lack SleB, presumably because CwlJ cannot be activated in the absence of Ca^{2+}-DPA (Black et al., 2005). Core demineralization also decreases spore responsiveness to LP (Igura et al., 2003). In addition, inhibitors that block the germination of spores by nutrients likely occurs through their action on the nutrient germinant receptors (Cortezzo et al., 2004a), and these inhibitors also block spore germination by LP (Wuytack et al., 2000; Black et al., 2005). A *B. subtilis* strain has been isolated in which a mutation in a gene of heretofore unknown function slows nutrient, dodecylamine, and LP (and also EHP) germination of spores (Aersten et al., 2005). This protein, encoded by *ykvU*, is likely a sporulation-specific membrane protein. However, the specific function of YkvU in the spore or even in sporulation is not known.

The major unanswered question about LP spore germination concerns the mechanism whereby LP activates the nutrient germinant receptors. HP can have effects on both proteins and membranes (Braganza and Worcester, 1986; Bartlett, 2002). Does LP act on the spore's inner membrane, the nutrient germinant receptor proteins, or on both? Are there other proteins, such as YkvU, that may be involved in this effect of LP? The major bar to the understanding of the mechanism of LP activation of the spore's germinant receptors is, of course, that we currently do not understand how these receptors work in nutrient germination. This is clearly a question that deserves intensive study.

EHP Germination

In contrast to LP germination, EHP does not trigger germination via activation of the nutrient receptors (Figure 2.6).

B. subtilis spores lacking all functional nutrient receptors germinate as well as wild-type spores with 500–600 MPa of pressure (Wuytack et al., 2000; Paidhungat et al., 2002; Black et al., 2007). In addition, a mutation blocking lipidation of the C-protein component of nutrient germinant receptors has no effect on EHP germination (Black et al., 2007) (Figure 2.6). GerD is also not required for EHP spore germination, and most inhibitors of nutrient receptor function (Cortezzo et al., 2004a) do not block the process, although some (such as Hg^{2+}) do (Wuytack

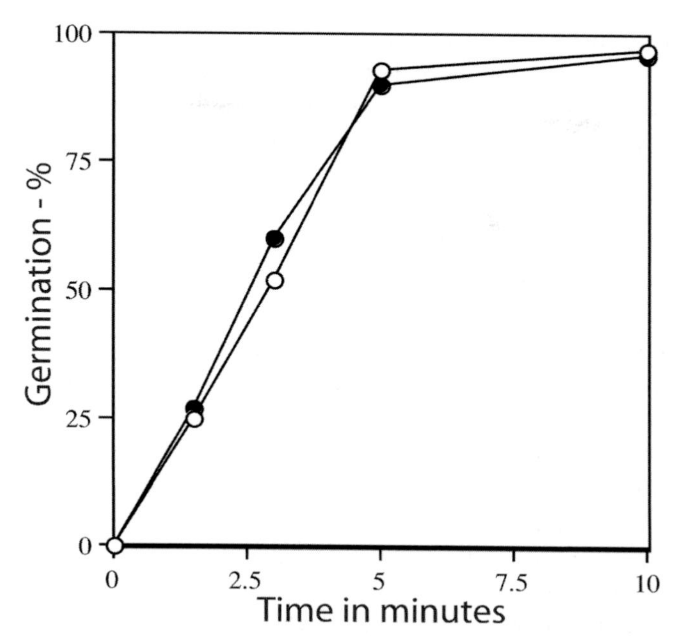

Figure 2.6. EHP germination of spores with or without nutrient germinant receptors. Spores of the isogenic *B. subtilis* strains PS533 (has all nutrient germinant receptors) or FB72 (lacks all functional nutrient germinant receptors (Paidhungat and Setlow, 2000) at an optical density of 1 were treated with a pressure of 500 MPa at 50°C in 50 mM Tris-HCl (pH 7.5). The percentage of spore germination was determined as described in the legend to figure 2.4. The symbols used are open circles (o) to designate PS533 spores, and filled circles (•) to represent FB72 spores. The data in this figure are from Black et al., 2007.

et al., 2000; Black et al., 2007). EHP does not directly activate CLEs, as *cwlJ sleB B. subtilis* spores release Ca^{2+}-DPA when treated with 500 MPa at a moderate temperature. However, either CLE is required for EHP-treated spores to complete spore germination (Paidhungat et al., 2002; Black et al., 2007).

Although there are some commonalities between the nutrient germination pathway and that for EHP germination, there are some differences. Spores germinated by EHP have undergone little SASP degradation, accumulate little ATP, and some resistance properties, in particular UV resistance, are retained (Wuytack et al., 1998). However, Ca^{2+}-DPA is released in EHP spore germination, even in the absence of CwlJ and SleB (Paidhungat et al., 2002; Black et al., 2007). Ca^{2+}-DPA is also released when *B. subtilis* spores are treated with EHP at elevated temperatures where it is likely that CLEs do not act (Knorr, 1999; Margosch et al., 2004; Black et al., 2007). These findings suggest that EHP germinates spores by causing Ca^{2+}-DPA release, and that the Ca^{2+}-DPA release triggers further events in germination. Indeed, EHP does not germinate DPA-less spores (Black et al., 2007). An EHP germinated spore will initially be in stage I of germination. This stage I germinated spore would be expected to rapidly progress through stage II, as is the case in nutrient germination. However, the stage I germinated spore generated by EHP does not progress rapidly to stage II of germination, although does so eventually (Wuytack et al., 1998). This slow progression through stage II of the EHP germinated spore is not understood. Possible explanations for this anomaly include (i) prevention of CwlJ action by EHP itself; or (ii) inactivation of CwlJ and/or SleB during EHP treatment. While EHP treatment does not rapidly give complete spore germination, as the treated spores likely "pause" after stage I, the stage I germinated spore has lost much resistance to wet heat as noted above, and as has been observed (Wuytack et al., 1998; Setlow et al., 2001). Thus the slow complete germination upon EHP treatment does not preclude spore inactivation.

While EHP germinates spores by causing Ca^{2+}-DPA release, the reason for the Ca^{2+}-DPA release is not clear. Does EHP open a normal Ca^{2+}-DPA channel in the spore's inner membrane, or does EHP alter the inner membrane properties such that Ca^{2+}-DPA can leak out? Which of these possible mechanisms is the right one is not clear, but it appears more likely that EHP directly opens channels or pores in the inner membrane. Ca^{2+}-DPA release by EHP takes place even at elevated temperatures where one might expect protein channels to be non-functional. In addition, a temperature sensitive mutation in a *spoVA* cistron, encoding a likely candidate for a Ca^{2+}-DPA channel protein, does not make EHP spore germination temperature sensitive (Wei et al., 2005). If EHP acts on the spore's inner membrane, then inner membrane fluidity might affect the spore's sensitivity to EHP. While there is no direct proof of this latter contention, spores grown at lower temperatures germinate more readily with EHP than do spores produced at higher temperatures, as shown with spores of several *Bacillus* species (Raso et al., 1998a, 1998b; Black et al., 2007). However, as found with LP germination, spore germination by EHP is unaffected by the level of unsaturated fatty acids in or by oxidative damage to the spore's inner membrane

(Black et al., 2007). In some regards the germination of spores by EHP resembles germination by dodecylamine. However, dodecylamine inhibits rather than accelerates spore germination by EHP (Black et al., 2007).

Conclusions and Unanswered Questions

Spore inactivation by HP is initiated by spore germination, giving rise to a more readily inactivated germinated spore that has completed either stage I or stage II of germination. The process is efficient, as $\geq$ 99% of a population of spores can be germinated by HP. However, processing of some foodstuffs requires much higher levels of spore inactivation, and promotion of these high levels of germination of spores of some species, such as *C. botulinum*, may not be possible. A major unanswered question in this regard is the reason that a small percentage of a population of spores are not germinated by HP. These latter spores seem likely to be the same super-dormant spores that are slow to respond to nutrient germinants. Unfortunately, the explanation for super-dormancy of spores is not known.

The action of LP and EHP in triggering germination is different, in that LP acts through the spore's nutrient germinant receptors, while EHP causes Ca^{2+}-DPA release. In both cases, these pressures act at the spore's inner membrane, either on membrane proteins or the membrane itself, or on both. Of obvious relevance then is the likely novel structure of the spore's inner membrane, which is currently not understood. If HP triggers spore germination largely by acting on the spore's inner membrane, then knowledge of this membrane's structure may suggest chemicals or other treatments that might facilitate HP action on the inner membrane. The structure of the spore's nutrient germinant receptors and how these proteins function to trigger spore germination are also not understood. Since Ca^{2+}-DPA release is a crucial event in spore germination by HP, whether triggered by activation of nutrient germinant receptors or directly, this is another event whose mechanism needs more detailed understanding. Clearly, while HP germination of spores holds significant promise in food processing, there is still much basic information that is lacking. Acquisition of this information should lead to a better understanding of HP germination, and may lead to improvements in spore inactivation by this treatment.

Acknowledgments

Work in the author's laboratory described in this chapter has been supported by grants from the National Institutes of Health (GM19598) and the Army Research Office to Peter Setlow, and from the U.S. Department of Agriculture to Dallas G. Hoover and Peter Setlow.

References

Ablett, S., A.H. Darke, P.J. Lillford, and D.R. Martin. 1999. Glass formation and dormancy in bacterial spores. *International Journal of Food Science and Technology* 34 (1):59–69.

Aersten, A., I. Van Opstal, S.C. Vanmuysen, E.Y. Wuytack, and C.W. Michiels. 2005. Screening for *Bacillus subtilis* mutants defective in pressure induced spore germination: Identification of *ykvU* as a novel germination gene. *FEMS Microbiology Letters* 243 (2):385–391.

Atluri, S., K. Ragkousi, D.E. Cortezzo, and P. Setlow. 2006. Co-operativity between different nutrient receptors in germination of spores of *Bacillus subtilis* and reduction of this co-operativity by alterations in the GerB receptor. *Journal of Bacteriology* 188 (1):28–36.

Bagyan, I., and P. Setlow. 2002. Localization of the cortex lytic enzyme CwlJ in spores of *Bacillus subtilis*. *Journal of Bacteriology* 184 (4):1289–1294.

Bartlett, D.H. 2002. Pressure effects on *in vivo* microbial processes. *Biochimica et Biophysica Acta* 1595 (1–2):367–381.

Bertsch, L.L., P.P. Bonsen, and A. Kornberg. 1969. Biochemical studies of bacterial sporulation and germination. XIV. Phospholipids in *Bacillus megaterium*. *Journal of Bacteriology* 98 (1):75–81.

Black, E.P., K. Koziol-Dube, D. Guan, J. Wei, B. Setlow, D.E. Cortezzo, D.G. Hoover, and P. Setlow. 2005. Factors influencing the germination of *Bacillus subtilis* spores via the activation of nutrient germinant receptors. *Applied and Environmental Microbiology* 71 (10):5879–5887.

Black, E.P., J. Wei, S. Atluri, D.E. Cortezzo, K. Koziol-Dube, D.G. Hoover, and P. Setlow. 2007. Analysis of factors influencing the rate of germination of spores of *Bacillus subtilis* by very high pressure. *Journal of Applied Microbiology* 102: 65–76.

Braganza, L.F., and D.L. Worcester. 1986. Structural changes in lipid bilayers and biological membranes caused by hydrostatic pressure. *Biochemistry* 25 (23):7484–7488.

Cano, R.J., and M. Borucki. 1995. Revival and identification of bacterial spores in 25- to 40-million-year-old Dominican amber. *Science* 268 (5215):1060–1064.

Chirakkal, H., M. O'Rourke, A. Atrih, S.J. Foster, and A. Moir. 2002. Analysis of spore cortex lytic enzymes and related proteins in *Bacillus subtilis* endospore germination. *Microbiology* 148 (8):2383–2392.

Cortezzo, D.E., K. Koziol-Dube, B. Setlow, and P. Setlow. 2004b. Treatment with oxidizing agents damages the inner membrane of spores of *Bacillus subtilis* and sensitizes the spores to subsequent stress. *Journal of Applied Microbiology* 97 (4):838–852.

Cortezzo, D.E., B. Setlow, and P. Setlow. 2004a. Analysis of the action of compounds that inhibit the germination of spores of *Bacillus* species. *Journal of Applied Microbiology* 96 (4):725–741.

Cortezzo, D.E., and P. Setlow. 2005. Analysis of factors influencing the sensitivity of spores of *Bacillus subtilis* to DNA damaging chemicals. *Journal of Applied Microbiology* 98 (13):606–617.

Cowan, A.E., D.E. Koppel, B. Setlow, and P. Setlow. 2003. A soluble protein is immobile in dormant spores of *Bacillus subtilis* but is mobile in germinated spores: Implications for spore dormancy. *Proceedings of the National Academy of Sciences of the United States of America* 100 (7):4209–4214.

Cowan, A.E., E.M. Olivastro, D.E. Koppel, C.A. Loshon, B. Setlow, and P. Setlow. 2004. Lipids in the inner membrane of dormant spores of *Bacillus* species are immobile. *Proceedings of the National Academy of Sciences of the United States of America* 101 (20):7733–7738.

Douki, T., B. Setlow, and P. Setlow. 2005. Photosensitization of DNA by dipicolinic acid, a major component of spores of *Bacillus* species. *Photochemical and Photobiological Sciences* 4(8):591–597.

Driks, A. 1999. The *Bacillus subtilis* spore coat. *Microbiology and Molecular Biology Reviews* 63 (1):1–20.

———. 2002a. Overview: Development in bacteria: Spore formation in *Bacillus subtilis*. *Cellular and Molecular Life Sciences* 59 (3):389–391.

———. 2002b. "Proteins of the spore core and coat." In: *Bacillus subtilis and Its Closest Relatives: From Genes to Cells*, ed. A.L. Sonenshein, J.A. Hoch, and R. Losick, pp. 527–535. Washington, DC: American Society for Microbiology.

Errington, J. 1993. *Bacillus subtilis* sporulation: Regulation of gene expression and control of morphogenesis. *Microbiology Reviews* 57 (1):1–33.

Gerhardt, P., and R.E. Marquis. 1989. "Spore thermoresistance mechanisms." In: *Regulation of Prokaryotic Development*, ed. I. Smith, R.A. Slepecky, and P. Setlow, pp. 43–63. Washington, DC: American Society for Microbiology.

Gould, G.W. 1969. "Germination." In: *The Bacterial Spore*, ed. G.W. Gould and A. Hurst, pp. 397–444. London, England: Academic Press.

Gould, G.W., and A.J.H. Sale. 1969. Initiation of germination of bacterial spores by hydrostatic pressure. *Journal of General Microbiology* 60 (3):335–346.

Igarashi, T., B. Setlow, M. Paidhungat, and P. Setlow. 2004. Analysis of the effects of a *gerF* (*lgt*) mutation on the germination of spores of *Bacillus subtilis*. *Journal of Bacteriology* 186 (10):2984–2991.

Igarashi, T., and P. Setlow. 2005. Interaction between individual protein components of the GerA and GerB nutrient receptors that trigger germination of *Bacillus subtilis* spores. *Journal of Bacteriology* 187 (7):2513–2518.

Igura, N., Y. Kamimura, M.S. Islam, M. Shimoda, and I. Hayakawa. 2003. Effects of minerals on resistance of *Bacillus subtilis* spores to heat and hydrostatic pressure. *Applied and Environmtenal Microbiology* 69 (10):6307–6310.

Jones, C.A., N.L. Padula, and P. Setlow. 2005. Effect of mechanical abrasion on the viability, disruption and germination of spores of *Bacillus subtilis*. *Journal of Applied Microbiology* 99 (6):1484–1494.

Kennedy, M.J., S.L. Reader, and L.M. Swierczynski. 1994. Preservation records of micro-organisms: Evidence of the tenacity of life. *Microbiology* 140 (10):2513–2529.

Kim, H., M. Hahn, P. Grabowski, D.C McPherson, M.M. Otte, R. Wang, C.C. Ferguson, P. Eichenberger, and A. Driks. 2006. The *Bacillus subtilis* spore coat protein interaction network. *Molecular Microbiology* 59 (2):487–502.

Knorr, D. 1999. Novel approaches in food-processing technology: New technologies for preserving foods and modifying function. *Current Opinion in Biotechnology* 10 (5):485–491.

Leuschner, R.G.K., and P.J. Lillford. 2003. Thermal properties of bacterial spores and biopolymers. *International Journal of Food Microbiology* 80 (2):131–143.

Makino, S., and R. Moriyama. 2002. Hydrolysis of cortex peptidoglycan during bacterial spore germination. *Medical Science Monitor* 8 (6):RA119–RA127.

Margosch, D., M.G. Gänzle, M.A. Ehrmann, and R.F. Vogel. 2004. Pressure inactivation of *Bacillus* endospores. *Applied and Environmental Microbiology* 70 (12):7321–7328.

Melly, E., P.C. Genest, M.E. Gilmore, S. Little, D.L. Popham, A. Driks, and P. Setlow. 2002. Analysis of the properties of spores of *Bacillus subtilis* prepared at different temperatures. *Journal of Applied Microbiology* 92 (6):1105–1115.

Moir, A., B.M. Corfe, and J. Behravan. 2002. Spore germination. *Cellular and Molecular Life Sciences* 59 (3):403–409.

Nicholson, W.L., N. Munakata, G. Horneck, H.J. Melosh, and P. Setlow. 2000. Resistance of *Bacillus* endospores to extreme terrestrial and extraterrestrial environments. *Microbiology and Molecular Biology Reviews* 64 (3):548–572.

Paidhungat, M., K. Ragkousi, and P. Setlow. 2001. Genetic requirements for induction of germination of spores of *Bacillus subtilis* by Ca^{2+}-dipicolinate. *Journal of Bacteriology* 183 (16):4886–4893.

Paidhungat, M., B. Setlow, W.B. Daniels, D. Hoover, E. Papafragkou, and P. Setlow. 2002. Mechanisms of initiation of germination of spores of *Bacillus subtilis* by pressure. *Applied and Environmental Microbiology* 68 (6):3172–3175.

Paidhungat, M., B. Setlow, A. Driks, and P. Setlow. 2000. Characterization of spores of *Bacillus subtilis* which lack dipicolinic acid. *Journal of Bacteriology* 182 (14):5505–5512.

Paidhungat M. and P. Setlow. 2000. Role of Ger-proteins in nutrient and non-nutrient triggering of spore germination in *Bacillus subtilis*. *Journal of Bacteriology* 182 (9):2513–2519.

———. 2002. "Spore germination and outgrowth." In: *Bacillus subtilis and Its Relatives: From Genes to Cells*, ed. J.A. Hoch, R. Losick, and A.L. Sonenshein, pp. 537–548. Washington, DC: American Society for Microbiology.

Popham, D.L. 2002. Specialized peptidoglycan of the bacterial endospore: The inner wall of the lockbox. *Cellular and Molecular Life Sciences* 59 (3):426–433.

Popham, D.L., J. Helin, C.E. Costello, and P. Setlow. 1996. Muramic lactam in peptidoglycan of *Bacillus subtilis* spores is required for spore outgrowth but not for spore dehydration or heat resistance. *Proceedings of the National Academy of Sciences of the United States of America* 93 (26):15405–15410.

Raso, J., and G. Barbosa-Canovas. 2003. Nonthermal preservation of foods using combined processing techniques. *Critical Reviews in Food Science Nutrition* 43 (8):265–285.

Raso, J., G. Barbosa-Canovas, and B.G. Swanson. 1998b. Sporulation temperature affects initiation of germination and inactivation by high hydrostatic pressure of *Bacillus cereus*. *Journal of Applied Microbiology* 85:17–24.

Raso, J., M.M. Gongora-Nieto, G.V. Barbosa-Canovas, and B.G. Swanson. 1998a. Influence of several environmental factors on the initiation of germination and inactivation of *Bacillus cereus* by high hydrostatic pressure. *International Journal of Food Microbiology* 44 (1–2):125–132.

Redmond, C., L.W. Baillie, S. Hibbs, A.J. Moir, and A. Moir. 2004. Identification of proteins in the exosporium of *Bacillus anthracis*. *Microbiology* 150 (2):355–363.

Rode, L.J., and J. W. Foster. 1960a. The action of surfactants on bacterial spores. *Archiv für Mikrobiologie* 36 (1):67–94.

———. 1960b. Mechanical germination of bacterial spores. *Proceedings of the National Academy of Sciences of the United States of America* 46 (1):118–128.

———. 1961. Germination of bacterial spores with alkyl primary amines. *Journal of Bacteriology* 81 (5):768–779.

Sale, A.J.H., G.W. Gould, and W.A. Hamilton. 1970. Inactivation of bacterial spores by hydrostatic pressure. *Journal of General Microbiology* 60 (3):323–334.

Setlow, B., S. Atluri, R. Kitchel, K. Koziol-Dube, and P. Setlow. 2006. Role of dipicolinic acid in resistance and stability of spores of *Bacillus subtilis* with or without DNA-protective δ-type small, acid-soluble proteins. *Journal of Bacteriology* 188 (11):3740–3747.

Setlow, B., A.E. Cowan, and P. Setlow. 2003. Germination of spores of *Bacillus subtilis* with dodecylamine. *Journal of Applied Microbiology* 95 (3):637–648.

Setlow, B., E. Melly, and P. Setlow. 2001. Properties of spores of *Bacillus subtilis* blocked at an intermediate stage in spore germination. *Journal of Bacteriology* 183 (16):4894–4899.

Setlow, P. 1994. Mechanisms which contribute to the long-term survival of spores of *Bacillus* species. *Journal of Applied Bacteriology* 76 (Symposium Supplement):49S–60S.

———. 1995. Mechanisms for the prevention of damage to the DNA in spores of *Bacillus* species. *Annual Review of Microbiology* 49:29–54.

———. 2001. Resistance of spores of *Bacillus* species to ultraviolet light. *Environmental and Molecular Mutagenesis* 38 (2):97–104.

———. 2003. Spore germination. *Current Opinion in Microbiology* 6 (6):550–556.

———. 2006. Spores of *Bacillus subtilis*: Their resistance to and killing by radiation, heat and chemicals. *Journal of Applied Microbiology* 101 (3):514–525.

Tovar-Rojo, F., M. Chander, B. Setlow, and P. Setlow. 2002. The products of the *spoVA* operon are involved in dipicolinic acid uptake into developing spores of *Bacillus subtilis*. *Journal of Bacteriology* 184 (2):584–587.

Vepachedu, V.R., and P. Setlow. 2004. Analysis of the germination of spores of *Bacillus subtilis* with temperature sensitive *spo* mutations in the *spoVA* operon. *FEMS Microbiology Letters* 239 (1):71–77.

———. 2005. Localization of SpoVAD to the inner membrane of spores of *Bacillus subtilis*. *Journal of Bacteriology* 187 (16):5677–5682.

Vreeland, R.H., W.D. Rosenzweig, and D.W. Powers. 2000. Isolation of a 250 million-year old halotolerant bacterium from a primary salt crystal. *Nature* 407 (6806):897–900.

Warth, A.D. 1978. Relationship between the heat resistance of spores and the optimum and maximum growth temperatures of *Bacillus* species. *Journal of Bacteriology* 134 (3):699–705.

———. 1980. Heat stability of *Bacillus cereus* enzymes within spores and in extracts. *Journal of Bacteriology* 143 (1):27–34.

Wei, J., V.R. Vepachedu, D.G. Hoover, and P. Setlow. 2005. Unpublished results.

Westphal, A.J., B.P. Price, T.J. Leighton, and K.E. Wheeler. 2003. Kinetics of size changes of individual *Bacillus thuringiensis* spores in response to changes in relative humidity. *Proceedings of the National Academy of Sciences of the United States of America* 100 (6):3461–3466.

Wuytack, E.Y., S. Boven, and C.W. Michiels. 1998. Comparative study of pressure-induced germination of *Bacillus subtilis* spores at low and high pressure. *Applied and Environmental Microbiology* 64 (9):3220–3224.

Wuytack, E.Y., J. Soons, F. Poschet, and C.W. Michiels. 2000. Comparative study of pressure- and nutrient-induced germination of *Bacillus subtilis* spores. *Applied and Environmental Microbiology* 66 (1):257–261.

Chapter 3

Inactivation of *Bacillus cereus* by High Hydrostatic Pressure

Murad A. Al-Holy, Mengshi Lin, and Barbara A. Rasco

Introduction

Bacillus cereus is a Gram-positive, sporeforming, rod-shaped, facultative anaerobic microorganism. *B. cereus* is capable of producing foodborne illnesses upon the ingestion of toxin or the vegetative form of the microorganism that can produce toxin in situ. *B. cereus* is ubiquitous in the environment and has been detected in various food systems such as milk, vegetables, potatoes, rice and other grains, and cereal foods such as batter, mixes and breadings, spices, and various sauces.

B. cereus produces two types of toxins, either of which can be produced at room or refrigeration temperatures. These toxins could be of an enterotoxigenic or an emetic nature and differ widely in terms of their resistance to physical and chemical factors. The extracellular diarrhogenic enterotoxin can be inactivated with relatively mild heat treatment (56°C, for 5 min) and proteases, whereas the emetic type toxin is extremely heat and pH resistant (Shinagawa et al., 1991). The number of *B. cereus* spores or vegetative cells necessary for causing enteric intoxication varies between 5 and 7 log cfu/g of food. High populations of *B. cereus* ($> 10^5$ per gram) contaminating a food may cause intoxication (Johnson, 1984). However, some strains of *B. cereus* might be capable of producing intoxication at counts as low as 3–4 log cfu/g of food (Granum and Lund, 1997).

B. cereus spores exhibit a very high heat resistance as manifested by their relatively high D-value ($D_{100} = 2–5.4$ min) (Choma et al., 2000).

These bacterial spores are also resistant to different physical treatments such as heat, high hydrostatic pressure (HHP), and pulsed electric fields, and to chemical agents such as sanitizers. Spores are resistant to these lethal agents because the protoplast, an internal component of the bacterial spore, is protected by a cortex composed of peptidoglycan and is further surrounded by a proteinacious coat (Smelt et al., 2002). Additionally, the very high resistance of *B. cereus* to preservation stresses is attributed to several other factors such as the presence of a calcium dipicolinate complex. Other components, called small acid soluble proteins or SASPs, prevent denaturation of the spore DNA. Also, the extremely low water content of the spore makes it remarkably more resistant to different environmental and preservation stresses (Jay, 2000). However, several additional factors influence spore resistance to HHP treatment, such as sporulation conditions and pressurization temperatures, pH, and water activity (Raso et al., 1998a, 1998b).

Few studies have been undertaken to investigate the effect of HHP treatment on the inactivation of *B. cereus* spores in foods or in model systems. This chapter summarizes the current research on the effect of HHP treatment on the inactivation of *B. cereus* in foods or model systems in relation to the factors affecting the pressure resistance and sensitivity of *B. cereus* to HHP treatment.

Mode of Inactivation of *B. cereus* by HHP Treatment

HHP treatment is generally considered to act on bacterial cell membranes and impair their permeability and ion exchange (Hoover et al., 1989). Microorganisms vary widely in their resistance to HHP treatment. Most often, bacterial vegetative cells are inactivated at pressures around 300–400 MPa at ambient temperature, whereas bacterial spores resist HHP treatment (Knorr, 1995). The inactivation of spores by HHP treatment can involve a two-step process: first, germination of the spores, and second, subsequent inactivation of the germinated spores (Fujii et al., 2002). Germination is the process by which a stimulus is applied to induce the dormant spores to convert to a metabolically active vegetative state. A variety of chemical and physical stimuli are known to induce germination. Heat shock is probably the most common stimulus. With heat shock, spores may require a period of heating at quite high temperatures (80°C for 15 min) to induce germination. Upon spore germination, spores lose their extraordinary resistance to different physicochemical agents. In principle, if all of the spores present in a food material could be induced to germinate, the food material could then be sterilized by a subsequent preservation treatment that would be milder than the treatment needed to inactivate ungerminated spores (Raso et al., 1998b). The germination process involves the release of dipicolinic acid (DPA), which is normally not present in vegetative bacteria. HHP between 50 and 300 MPa induces spore activation and germination, and the inactivation of the germinated spores proceeds faster at neutral pH. Therefore, HHP treatments can be devised to preserve low-acid foods by inducing spore germination and subsequent inactivation of the germinated spores (Smelt, 1998).

Effect of Pressure Intensity

The susceptibility of different strains of *B. cereus* to HHP treatment is shown in Table 3.1. Different strains of *B. cereus* show little variability in terms of their susceptibility to HHP treatment regardless of the intensity of the HHP applied. No more than 1 log unit difference in the inactivation of the tested *B. cereus* strains was observed, with an approximate 6 log unit reduction resulting from a treatment of 500 MPa/60°C for 30 min or of 200 MPa/45°C for 30 min followed by heat treatment at 60°C for 10 min (Van Opstal et al., 2004).

Pressure-induced spore inactivation is usually preceded by spore germination events. Researchers have noted that different *Bacillus* spp. spores can germinate more efficiently at pressures between 200 and 500 MPa (Nakayama et al., 1996). *B. cereus* in particular can be germinated optimally at about 250 MPa HHP (Raso et al., 1998b). Pressure-induced spore germination does not depend upon the level of pressure applied. Comparable levels of germinated *B. subtilis* spores were obtained at 100 MPa and 500 MPa (Wuytack et al., 1998). However, the germinated spores differed considerably in their sensitivity to inactivation by HHP treatment. *B. subtilis* spores germinated at 100 MPa were more sensitive to HHP inactivation compared to spores germinated at 500 MPa. However, following germination, the spores exhibited equivalent levels of sensitivity to heat treatment regardless of the germination pressure applied. Notwithstanding, spores germinated at 100 MPa were more sensitive to inactivation by either hydrogen peroxide or UV light

Table 3.1. High pressure inactivation of four *Bacillus cereus* strains in milk

		Viability Reduction (Log N_0/N)) Pressure Treatment Time	
Bacillus cereus Strain	Treatment	15 min	30 min
LMG6910	A	5.4	6.7
	B	5.1	7.4
INRAAV TZ415	A	6.6	6.7
	B	5.4	6.3
INRAAV P21S	A	5.6	5.8
	B	6.8	7.2
INRAAV Z4222	A	5.6	6.2
	B	6.1	6.4

A: Pressure treatment at 500 MPa/60°C.
B: Pressure treatment of 200 MPa, 45°C, followed by heat treatment at 60°C for 10 min.
N_0: Represents the plate count of the untreated spore suspension (6×10^6 spore/ml), and N represents the plate count after high pressure treatment.
(Table adapted from Van Opstal et al., 2004.)

compared to spores germinated at 600 MPa. This lack of sensitivity to inactivation may be attributed to the lack of degradation of the (SASPs) at high pressure (600 MPa).

In one of the few experiments conducted in a food system, less than 1 log unit of *B. cereus spores* were inactivated in pork slurries by HHP treatment of 600 MPa, and this pressure treatment caused coagulation and discoloration of the product. However, other vegetative pathogenic bacteria were inactivated at lower pressures (300–400 MPa) in pork slurries (Shigehisa et al., 1991). Also, using HHP treatment to process tofu at 400 MPa at 5°C for 45 min decreased the *B. cereus* spore count by only 1 log unit. This limited inactivation capacity was attributed to the composition of tofu that may provide a baro-resistant effect of the bacterial spores (Prestamo et al., 2000).

HHP treatment can cause sublethal injury to microbes in pressure-treated foods (Yuste et al., 2004). Injured microbial cells have the potential to resuscitate and resume growth in conditions favorable for growth, such as plentiful nutrient availability and suitable temperatures for prolonged storage times. These conditions are commonly found in foods. This aspect is important to consider, especially for pasteurized and refrigerated foods, because *B. cereus* is psychrotropic and can survive and grow at refrigeration temperatures.

Comparison of Pressure Resistance and Heat Resistance of *B. cereus*

The growth stage of bacterial cells greatly affects their sensitivity to HHP inactivation. Bacterial cells in the stationary phase are considerably more resistant to different preservation treatments, including HHP treatment, compared to bacterial cells in the exponential phase of growth (Knorr, 1995). *B. cereus* cells in the stationary phase showed less susceptibility to HHP inactivation compared to exponential phase cells when treated with 400 MPa at either 8 or 30°C (McClements et al., 2001). It was reported that changes in the cell membrane fatty acid composition of *Escherichia coli* are responsible for the increased stationary phase resistance to heat treatment (Katsui et al., 1981). The same factors have been implicated in causing the increasing pressure resistance of bacteria in the stationary phase. However, the mechanisms of resistance to heat and to pressure are probably quite different.

Nakayama et al. (1996) compared the heat and pressure resistance of six strains of *Bacillus* spores and reported that *B. stearothermophilus*, an extremely heat resistant strain, was the most pressure-labile. Conversely, *B. megaterium*, a relatively heat sensitive strain, was not inactivated by HHP treatment of 1,000 MPa for 400 minutes. These findings indicate that there is not necessarily a correlation between pressure resistance and heat resistance of bacteria, at least for *Bacillus* spp. Similarly, the endospore former *Thermoanaerobacterium thermosaccharolyticum* is ten times more heat resistant than is *Clostridium botulinum*. Yet HHP treatment for 4 min at 80°C for both species resulted in >4.5 log units reduction of *T. thermosaccharolyticum* compared to <2.5 log units reduction of *C. botulinum* (Margosch et al., 2004).

It is worthwhile to mention that high sporulation temperatures are associated with increased heat resistance of *B. cereus* spores (Setlow, 1994). In comparison, high sporulation temperatures were accompanied by increased lethality of bacterial spores by HHP treatment. This observation would suggest that the factors associated with the heat resistance of bacterial spores derived from sporulation at high temperatures do not similarly impart *B. cereus* spores with resistance to inactivation by HHP.

Margosch et al. (2004) hypothesized that the combined effect of HHP treatment and heat on the inactivation of bacterial spores is due to DPA release but not due to the pressure sensitivity of spores during HHP-induced spore germination. Also, the prominent ability of some bacterial endospores to resist HHP inactivation may be due to their outstanding ability to retain DPA during HHP treatment.

Table 3.2 compares the heat and pressure resistance of five different *Bacillus* spp. as indicated by their D-values (the time required to achieve 90% inactivation of the spore population under the conditions specified). *Geobacillus stearothermophilus* spores were the most pressure-resistant bacterial spores, whereas *B. cereus* and *B. coagulans* were the most pressure-labile. The D-value for pressure resistance was almost eleven times higher for *G. stearothermophilus* compared to *B. cereus*. However, there was no correlation between heat resistance and pressure resistance for *Bacillus* spp. spores (Watanabe et al., 2003; Nakayama et al., 1996; Smelt et al., 2002), again suggesting that spore inactivation mechanisms by heat and pressure are different.

In recent studies, combined treatments of heating and pressure have been used for spore inactivation. There appears to be a synergistic effect of HHP treatment and heat on the inactivation of *B. anthracis* spores (Cléry-Barraud et al., 2004). For example, at 500 MPa, the survival time of *B. anthracis* dropped from about 160 min at 20°C to 19 min at 45°C and to 4 min at 75°C.

Mechanisms of HHP Resistance of *B. cereus*

The presence of a DPA-calcium complex exerts a protective effect on spores against different physical and chemical inactivation stresses, including HHP treatment. This complex plays a role in the protection of spore proteins against pressure-induced water penetration through the spore wall. Pressure-induced release of DPA-calcium complex leads to modifications in spore proteins and changes in membrane permeability coupled with rehydration of spore core (Smelt et al., 2002). Additionally, it has been shown that a pressure of 100 MPa induces spore germination by activation of germinant receptors, and a pressure of 550 MPa opens

Table 3.2. Effect of heat treatment at 85°C, pressure treatment at 200 MPa and 65°C, and CO_2 treatment at 35°C and 30 MPa on the resistance of different *Bacillus* spp. spores

Organism	Treatment	D-value (min)
Geobacillus stearothermophilus	Heat	
	Pressure	75.2
	CO_2	385.0
Bacillus subtilis	Heat	19.0
	Pressure	9.3
	CO_2	1667.0
Bacillus coagulans	Heat	9.5
	Pressure	6.9
	CO_2	164.0
Bacillus cereus	Heat	8.5
	Pressure	6.9
	CO_2	133.0
Bacillus licheniformis	Heat	7.9
	Pressure	8.5
	CO_2	182.0

(Adapted from Watanabe et al., 2003.)

channels for the release of the DPA-calcium complex from the spore core (Paidhungat et al., 2002).

Bacterial spores that have nutrient receptors can be activated by mild HHP treatment (100–300 MPa), hence inducing their germination. *B. subtilis* spores lacking nutrient receptors showed a limited and slow spore germination induced by HHP treatment at 150 MPa for 6 min (Black et al., 2005). This may explain why super-dormant (see the section "Super-Dormant *B. cereus* Spores") spores do not germinate when exposed to HHP treatment. The DPA-calcium complex normally comprises $\geq$20% of the dry weight of the spore core. Spores that lack an endogenous DPA-calcium complex do not germinate with moderate HHP treatment (150 MPa).

Wuytack et al. (1998) studied the resistance of *B. subtilis* spores to germination by HHP treatment and observed that the process of spore germination at 600 MPa was incomplete. These researchers surmised that this was because of the presence of SASPs that did not undergo degradation and hindered HHP-induced germination at 600 MPa. It was also observed that degradation takes place at mild pressure (100 MPa). SASPs are known to protect spore DNA not only against HHP treatment but also against hydrogen peroxide and UV-light exposure. A possible explanation for the limited SASP degradation under extreme HHP treatments was attributed to the inactivation of a germinative protease by HHP, which is an enzyme responsible for SASP degradation.

Again, as with heat treatment or any other process to inactivate microbes, after HHP processing of food, the absence of bacterial growth on a general or a selective media does not necessarily mean that all bacteria in this food has been eliminated. It is possible that some sublethally injured cells may recover if an appropriate enrichment step is applied and if proper conditions exist for growth in the treated food (Yuste et al., 2004).

Super-Dormant *B. cereus* Spores

Normally, after HHP-induced germination, a small fraction of super-dormant spores remains that resists inactivation by different treatments such as HHP, heat, hydrogen peroxide, or UV light. Super-dormant spores are those that remain ungerminated in a medium that induces the germination of a majority of the spore population (Raso et al., 1998b). One of the suggested reasons that HHP treatment cannot inactivate these super-dormant spores is because of the lack of core rehydration. Another possibility is that SASP degradation in a super-dormant spore is inhibited at high levels of HHP (500–600 MPa) and this, consequently, affects the spore's ability to germinate (Wuytack et al., 1998). Additionally, a subpopulation of super-dormant spores could include bacterial spores lacking nutrient receptors, which have a slow and limited spore germination induced by HHP treatment at 150 MPa for 6 min (Black et al., 2005). It has been hypothesized that HHP treatment could induce the development of mutant pressure-resistant spores (Cléry-Barraud et al., 2004).

Germination of the super-dormant spores must be accomplished if one is to ascertain how they can be inactivated. Inactivation of super-dormant spores is different than the rest of the spore population. For example, Cléry-Barraud et al. (2004) showed that survival curves for

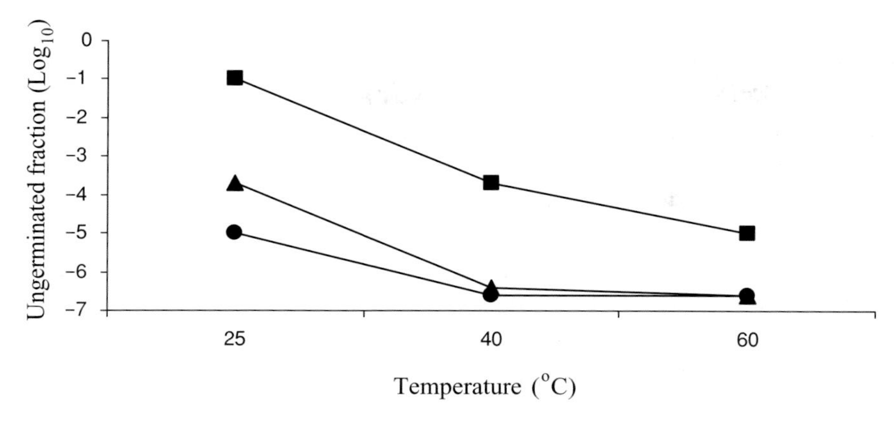

Figure 3.1. Influence of pressurization temperature on the germination of *Bacillus cereus* sporulated at 20°C (■), 30°C (▲), and 37°C (•) by high pressure at 400 MPa for 15 min. (Adapted from Raso et al., 1998a.)

HHP-treated *B. anthracis* exhibited first-order pressure-inactivation kinetics. However, *B. cereus* spores exhibited multiple order exponential inactivation kinetics attributable to the presence of super-dormant spores, which were resistant to prolonged HHP treatment at 500 MPa and 75°C. Others have shown that combined pressure and temperature treatments can induce germination of super-dormant spores, permitting inactivation kinetic models to be developed. Cycling pressure treatment between 60 and 500 MPa at a relatively high temperature (60°C) has been shown to favor the germination of super-dormant spores of *B. subtilis* (Sojka and Ludwig, 1997).

Effect of Sporulation and Pressurization Temperatures on *B. cereus*

Pressurization temperature is one of the most important factors affecting the susceptibility of bacterial spores to HHP treatment. HHP-induced germination of *B. cereus* spores proceeded at a much higher rate when a pressurization temperature of 30°C at 400 MPa was used instead of 8°C (McClements et al., 2001). Also, the extent of germination is influenced by sporulation temperature (Raso et al., 1998a, 1998b). The fraction of *B. cereus* spores that undergo germination by HHP increases with increasing sporulation temperature (Figure 3.1). An increase in the sporulation temperature from 20°C to 37°C led to greatly enhanced spore germination up to a certain point. There is an optimal level beyond which inactivation begins, and this trend is more pronounced at higher pressure levels. HHP treatments of 250 MPa at 25°C for 15 min induced spore germination regardless of the medium pH (3.5–7.8). Yet, little *B. cereus* spore inactivation was detected in the same treatment (Raso et al., 1998b).

Sporulation temperature greatly affects the extent of HHP-induced germination. Raso et al. (1998a) pointed out that at low sporulation temperature (20°C), *B. cereus* spores showed more resistance to spore germination by HHP treatment compared to sporulation temperatures of 30 and 37°C. The same pattern is true for the pressurization temperature. *B. cereus* spores show considerable ability to germinate when using relatively high pressurization temperatures (40–60°C) compared to pressurization at the relatively low temperature of 25°C (Figure 3.1). At higher temperatures, *B. cereus* spore germination proceeds at a faster rate as pressure intensity increases. At 25°C spores germination was optimum at 250 MPa (Raso et al., 1998a).

Since the inactivation of bacterial spores should be preceded by spore germination, most often HHP-induced spore germination is coupled with spore inactivation by an HHP treatment. This is effective for *B. cereus* only at relatively high pressures ($\geq$ 400 MPa). Low pressurization temperatures are often accompanied by lower spore inactivation compared to the extent of inactivation seen at higher pressurization temperatures $\geq$40°C (Figure 3.2). At higher temperatures, the sensitivity of *B. cereus* spores to HHP inactivation increases probably because germinated spores are inactivated directly by high temperature (~60°C) (Raso et al., 1998a). Nonetheless, it is extremely difficult to pressure inactivate all spores present in a certain medium. This is likely due to the heterogeneity of the *B. cereus* spore population and the presence of a super-dormant group of spores, which remain ungerminated even after the application of germinants or treatments that are otherwise effective.

The rate of germination is also temperature dependent and greater at higher temperatures. Generally, the rate of pressure-induced germination of *B. subtilis* spores prepared at 44°C was almost 10-fold higher than that of spores prepared at 23°C. However, this observation is not always accurate. For example, the rate of germination by HHP treatment in the presence of L-alanine decreased by 1.5- to 2.0-fold in *B. subtilis*

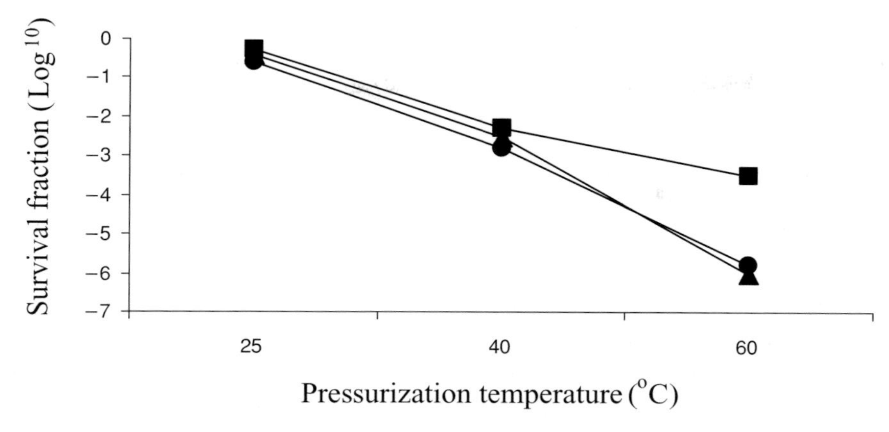

Figure 3.2. Influence of pressurization temperature on the inactivation of *Bacillus cereus* sporulated at 20°C (■), 30°C (▲), and 37°C (•) by high pressure processing at 400 MPa for 15 min. (Adapted from Raso et al., 1998a.)

spores prepared at 44°C compared to 23°C (Black et al., 2005). One explanation for the enhanced HHP-induced spore germination with increasing sporulation temperature is that activity of nutrient receptors increases at higher sporulation temperatures (Black et al., 2005).

Effect of Growth Medium or Food Source

Medium Constituents

The extent of pressure-induced germination of bacterial spores is highly variable and depends upon several factors such as the organism tested, the process conditions evaluated, and the physicochemical environment (Raso et al., 1998a, 1998b). The nature of the growth medium significantly affects the extent of *B. cereus* spore germination. For instance, regardless of the level of HHP applied, almost complete germination of the spores was obtained in skim milk. In comparison, in a phosphate buffer, a material of similar pH but lacking in protein and amino acids, the degree of *B. cereus* spore germination was markedly less and increased with increasing pressure from 100–600 MPa at 40°C (Figure 3.3; see also Van Opstal et al., 2004). This difference may be due to the presence of certain amino acids such as L-alanine, L-leucine, and L-isoleucine that induce spore germination and enhance *B. cereus*

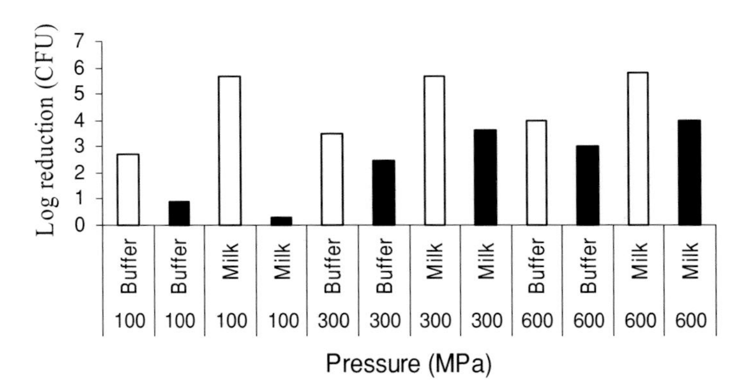

Figure 3.3. Germination (open bars) and inactivation (filled bars) of *Bacillus cereus* LMG6910 spores by pressure treatment for 30 min at 40°C in potassium phosphate buffer (100 mM, pH 6.7) and milk. The phosphate buffer and milk were inoculated by 6×10^6 spore/mL. (Adapted from Van Opstal et al., 2004.)

growth (Agata et al., 1999; Raso et al., 1998b). Other milk components such as fat appear to have little influence on the initiation of germination and on a subsequent inactivation treatment of *B. cereus* spores by HHP (Raso et al., 1998b). However, the effect of L-alanine was pressure dependent with the presence of L-alanine in a medium increasing both the germination and subsequent inactivation of *B. cereus* at 250 MPa, but not at 690 MPa. This effect was observed at different temperatures over the range of 25–37°C (Table 3.3).

Several studies have been undertaken to examine microbial inactivation of different types of microorganisms by HHP treatment in buffers or laboratory media (Raso et al., 1998a, 1998b; Fujii et al., 2002; Oh and Moon, 2003; Watanabe et al., 2003). However, the results of such studies cannot be extrapolated to real foods. The intrinsic factors of certain foods play an important role in determining the microbiological stability and safety of that food. Some food factors such as acidity, water activity, and the presence of anti-microbial constituents influence the efficiency of HHP treatment inactivation of a particular microorganism. Additionally, these food components may affect food stability after HHP treatment by enhancing the HHP-preservation effect or by stimulating the pressure resistance or by promoting injured microorganisms to resuscitate and grow back to levels equivalent to that found before the original HHP application (Smelt, 1998). Milk and cream,

Table 3.3. Influence of the presence of L-alanine on the initiation of germination and inactivation of *Bacillus cereus* in McIlvaine citrate phosphate buffer by high pressure treatment. Initial spore count ~1×10^7 spore/mL

		Germinated spores (Log_{10})				Inactivated spores (Log_{10})			
		L-alanine (mM)				L-alanine (mM)			
Sporulation Temperature (°C)	Treatment	0	1	10	100	0	1	10	100
20	250 MPa, 25°C, 15 min	1.40	1.39	1.39	1.78	0.25	0.25	0.22	0.39
	690 MPa, 40°C, 2 min	3.30	3.30	3.40	3.48	1.95	2.03	1.78	2.30
37	250 MPa, 25°C, 15 min	3.20	3.57	4.11	4.22	0.39	0.41	0.43	0.50
	690 MPa, 40°C, 2 min	6.90	7.00	7.00	7.00	5.00	5.12	5.12	5.20

(Adapted from Raso et al., 1998b.)

for instance, provide a baroprotective effect to vegetative pathogenic microorganisms (Patterson et al., 1995). Skim milk, on the other hand, enhances the HHP-inhibitory effect against spores of *B. cereus*, probably because skim milk contains some amino acids such as L-alanine that induce spore germination and hence destruction of *B. cereus* germinated spores (Van Opstal et al., 2004).

Besides amino acids, other food components can alter the effectiveness of a high pressure treatment. Shearer et al. (2000) demonstrated that the antimicrobial effect of sucrose laurate against *B. cereus* spores could be enhanced when combined with HHP treatment. The combination of sucrose laurate with pressure was more inhibitory than sucrose laurate alone or sucrose laurate combined with mild heat. However, using HHP (392 MPa for 10 min) combined with heating at 45°C in the presence of sucrose laurate (0.01%) resulted in about 5 log units reduction of *B. cereus* spores in a beef puree (baby food). Compared to *B. cereus*, the combination of HHP (392 MPa for 10 min), and heating at 45°C in the presence of 0.1% sucrose laurate was less effective against *B. subtilis*, where this combination resulted in about a 3 log units reduction (Shearer et al., 2000).

The addition of sucrose laurate following HHP treatment resulted in the same inhibitory effect as compared to both of them when combined simultaneously. It was hypothesized that the inhibitory effect of sucrose laurate is enhanced by HHP treatment because HHP treatment helps to deposit sucrose laurate on the spore coat and thereby change the surface hydrophobicity and water permeability of the spore during pressurization.

Acidity

The effect of sporulation pH and suspension pH on the inactivation of *B. cereus* by HHP treatment was studied by Oh and Moon (2003). *B. cereus* spores obtained from a medium with a pH = 6 exhibited higher resistance to HHP (≥ 300 MPa) at temperatures ≥ 40°C compared to treatments at the same pressure at pH 7 and 8. Nonetheless, the effect of sporulation pH was minor when the pressurization temperature was 20°C, regardless of the pressure applied over the range of 0–600 MPa. Therefore, the use of a pH-temperature combination may play an important concerted role in discerning the sensitivity of *B. cereus* spores to HHP treatment, but only under certain conditions. The effect of

Table 3.4. Effect of pH (4.5–8.0) at different pressures and pressurization temperatures on the inactivation of *Bacillus cereus* spores*

		Reduction in *Bacillus cereus* Spores (Log CFU)			
		pH	pH	pH	pH
Pressure (MPa)	Temperature (°C)	4.5	6.0	7.0	8.0
450	20	0.0	1.8	2.0	1.9
	40	3.4	5.0	5.8	4.2
	60	4.0	5.8	6.3	5.8
600	20	1.3	3.2	2.3	2.4
	40	4.0	5.5	6.4	5.3
	60	4.5	7.0	7.4	6.6

* *B. cereus* spores were inoculated in McIlvaine buffer at a concentration of 10^8–10^9 spores/mL. (Adapted from Oh and Moon, 2003).

sporulation pH on the inactivation of *B. cereus* is maximal when pressure is applied at temperature $\geq$ 40°C. This phenomenon is apparently related to spore germination, as a prerequisite step for spore inactivation. Moderate pressures (300 MPa) and pressurization temperatures of $\geq$ 40°C are required to induce spore germination (Oh and Moon, 2003).

The effect of suspension pH on the inactivation of *B. cereus* spores by HHP treatment is shown in Table 3.4. Spores suspended in a lower pH medium (4.5) tend to exhibit more resistance to HHP inactivation compared to suspension with pH $\geq$6.0 at 450 MPa and 600 MPa. Additionally, the extent of inactivation is dependent upon the pressurization temperature as well as on the intensity of the pressure applied. Nonetheless, not all *Bacillus* spp. respond in the same manner to pH; for example, *B. coagulans* spores showed greater sensitivity to pressurization at a low pH than other *Bacillus* spores (Roberts and Hoover, 1996).

The prominent resistance of *B. cereus* spores to HHP treatment at low pH (4.5) may be attributed to the limited germination of *B. cereus* spores that takes place at this pH. It may also be a result of a possible change in electrochemical charges of some structural components such as calcium dipicolinate that takes place during spore formation (Oh and Moon, 2003). Raso et al. (1998a) reported that HHP-induced germination of bacterial spores depends upon the pH of the suspension buffer. *B. cereus* suspended in a buffer of a pH = 4.5 showed substantially higher resistance to HHP-induced germination than suspensions at pH $\geq$ 6.0 at the same pressure and temperatures (Oh and Moon, 2003).

Pressure-induced spore germination is highly dependent on the pH of the treatment medium. Increasing the medium pH from mildly acidic to neutral induces more germination of the *B. cereus* spores. The maximum germination of *B. cereus* spores was obtained around a neutral pH (pH = 6.0). However, the influence of pH on spore germination is highly dependent upon pressurization temperature. For instance, the pH slightly affected spore germination at pressurization temperature of 20°C. Raso et al. (1998a) indicated that optimal spore germination of *B. cereus* was obtained at temperatures between 30 and 37°C and at a pH around neutral at different pressure levels. On the contrary, this is not generally true for all *Bacillus* spp. For example *B. coagulans* spores showed increasing sensitivity to HHP treatments at lower pH values (Roberts and Hoover, 1996).

The role of pH may be tied with a spore's retention of DPA. Margosch et al. (2004) stated that thermal and HHP treatment resistance of bacterial spores was correlated with the spore's ability to retain DPA. The release of DPA from endospores increases as the pH of the medium or food decreases. Researchers have suggested that HHP treatment induces the opening of channels in *B. cereus* spores that result in the release of DPA, and this is subsequently followed by activation of the germination pathway (Paidhungat et al., 2002). Higher pressurization temperatures and low medium pH induce a greater release of DPA. Low pH causes a considerable increase in the permeability of the spore barrier permeability, which leads to a concomitant release of DPA and hydration of the spore core, hence increasing a spore's HHP sensitivity (Setlow et al., 2002).

In comparison, the sensitivity of vegetative bacterial cells to HHP treatment is enhanced when these bacteria are present in a medium with an acidic pH. However, bacterial spores exhibit a remarkable resistance at neutral pH when HHP treatment is applied. After spore germination at pressures between 50 and 300 MPa, germinated spores become more sensitive to HHP inactivation at neutral pH. Furthermore, a suboptimal medium or food pH may pose inhibitory effects on the recovery of sublethally injured cells (Smelt, 1998).

Generally, the presence of organic acids increases a bacteria's sensitivity to a processing treatment, as organic acids generally possess antimicrobial activity in the protonated form. However, no specific effect of organic acids during pressure treatment has been observed to date. This may be due to the fact that pressure restricts the inhibitory effect of organic acids by possibly favoring ionization of organic compounds (Smelt et al., 2002).

Water Activity

Water activity is a crucial factor that affects the efficiency of HHP treatment against microorganisms. Water activity is critical in controlling germination of bacterial spores by HHP as well as at ambient pressure. At relatively low HHP-treatment levels, water activity has little effect on spore germination at low temperatures (20°C), and the extent of germination increases as the pressurization temperature increases. The inactivation of *B. cereus* spores at 250 MPa is negligible regardless of the pressurization temperature. In general, no *B. cereus* germination and inactivation have been observed at $a_w \leq 0.92$, regardless of the HHP level or pressurization temperature. The presence of a solute such as sucrose in the pressurization medium lowers the water activity and provides a baroprotective effect for *B. cereus* spores. The presence of sucrose in a medium may induce osmotic dehydration of the bacterial spore protoplast by lowering water activity. This may explain the high resistance of *B. cereus* spores to germination and inactivation by HHP treatment.

Generally, low water activity protects microbial cells against HHP treatment. Apart from the osmotic effect of low water activity induced by salt, carbohydrates, for example, provide a protective effect against HHP treatment. However, despite the fact that low water activity provides a baroprotective effect for microorganisms during HHP treatment, HHP injured bacterial cells are less likely to recover at suboptimal values of water activity (Smelt et al., 2002).

Alternating or Pulsed HHP Treatment

Applying successive but intermittent pressure cycles could be a prudent approach to inactivate *B. cereus* spores. Oscillatory pressure treatments

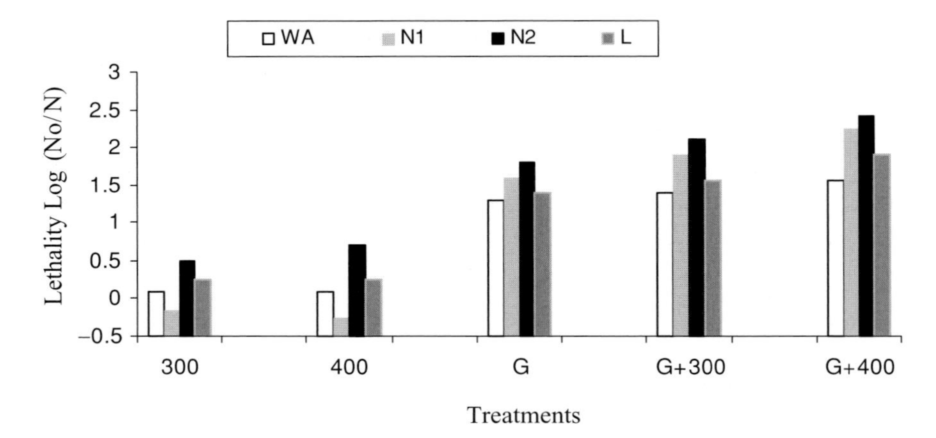

Figure 3.4. Lethality of *Bacillus cereus* in a model cheese treated with high hydrostatic pressure treatment (300–400 MPa, 30°C, 15 min) with or without additives or an HHP germination treatment. WA = without additives; N1 = 0.05 mg of nisin/L of milk; N2 = 1.56 mg of nisin/L of milk; L = 22.4 mg of lysozyme/L of milk; G = germination treatment of 60 MPa/210 min at 30°C. (Adapted from López-Pedemonte et al., 2003.)

for inactivating bacterial spores take advantage of the facts that pressures between 50 and 400 MPa can stimulate spores to germinate and that germinated spores have decreased resistance to pressures than do dormant spores. By alternating pressure treatment between 60 and 500 MPa by cycles of 1 min, spores of *B. subtilis* could be reduced by a factor of $>10^8$ (Wuytack et al., 1998).

The application of a single-step treatment of HHP (300 or 400 MPa) to model cheeses resulted in a limited inactivation (<1 log cycle) of *B. cereus* (Lopez-Pedemonte et al., 2003). The influence of the single HHP treatment (300 or 400 MPa) versus two cycles of HHP (60 + 300 MPa or 60 + 400 MPa) at 30°C was studied. The use of a low pressure (60 MPa) as a germinative step resulted in a more pronounced inactivation effect of the two-step treatment compared to the single-step treatment (Figure 3.4). The maximum inactivation obtained using 60 MPa + 400 MPa for 15 min at 30°C was 1.6 log units reduction (Lopez-Pedemonte et al., 2003). However, a two-step HHP treatment (60 MPa/210 min + 500 MPa/15 min at 40°C) in a model solution system resulted in a 6 log units reduction of *B. subtilis* spores (Sojka and Ludwig, 1994). The difference in the rate of inactivation may be attributed to the higher pressure and temperature in the latter study, in addition to the fact that actual food systems usually provide baroprotective effect for spores.

The HHP-induced germination of *B. cereus* spores to high levels may make it possible to produce safe milk with a reasonable shelf life by applying a preservation treatment subsequent to HHP treatment that would kill the germinated spores. Nonetheless, it seems difficult to kill all the germinated spores with a single step of HHP treatment at low (100 MPa) as well as at high (600 MPa) levels of HHP treatment (Van Opstal et al., 2004).

It has been noted that spores of *B. stearothermophilus*, when subjected to six pressure cycles of 600 MPa at 70°C for 5 min each, were reduced by 4–6 log units (Hayakawa et al., 1994). Additionally, alternating oscillatory pressure treatments between 60 and 500 MPa with holding times of 1 min could bring about more than an 8 log reduction in the number of *B. subtilis* spores under certain conditions (Smelt, 1998). Repeated applications of differing pressure cycles enhance HHP inactivation probably due to the dual effect of germination and subsequent inactivation by HHP treatment.

HHP Inactivation of *B. cereus* and Hurdle Technologies

The use of an intelligent combination of different food preservation stresses to secure the microbial, sensory, and nutritive stability of food has received a great deal of interest by the food industry (Leistner, 1992). HHP can be used to improve food safety and extend shelf life of foods by inhibiting microbial growth and inactivating food enzymes. In general, HHP treatment is effective at reducing most vegetative bacteria, yeasts, and molds at treatment pressures between 300 and 700 MPa (Smelt, 1998). In general, Gram-positive bacteria are more resistant to HHP than Gram-negative bacteria (Yuste et al., 2004). However, bacterial spores are much more resistant to environmental and food preservation stresses compared to their vegetative counterparts. Smelt (1998) stated that bacterial spores showed resistance to HHP treatment up to 1,000 MPa. This represents a great challenge to the food industry, where there is an intensive search for new technologies that aim to maintain the fresh-like quality attributes of food and keep food safe for human consumption for a reasonable period of time. Up to now, the most widely used method to kill bacterial spores in food is heat. However, the high temperatures used to inactivate spores cause great losses in nutrients, flavor, and texture of foods. Hence, HHP treatment emerges as a potential nonthermal, or more correctly, adiabatic food processing technique that preserves food without excessively degrading its quality (Martens and Knorr, 1992).

HHP has a unique ability to induce spore germination. The exceptional resistance of bacterial spores to various food preservation stresses such as heat, radiation, chemical preservatives, and HHP will be lost when bacterial spores germinate. Therefore, HHP-induced spore germination may be applicable as a potential hurdle that can be combined with other preservation methods to destroy bacterial spores in food.

High levels of HHP treatment (600 MPa) at slightly elevated temperatures (40–60°C) induced a moderately higher spore germination compared to lower pressure treatment (200 MPa/40°C). Higher pressurization temperatures (40–60°C) induced better spore germination compared to the lower temperature of 20°C (Oh and Moon, 2003).

A considerable reduction of *B. cereus* spores ($>$ 1 log unit) in milk is possible if HHP treatment is conducted at 400 MPa/60°C for 30 min. The application of a less severe, two-step HHP treatment leads to more efficient inactivation. More than 6 log units of inactivation of *B. cereus* spores was possible by applying an HHP treatment of 200 MPa/45°C followed by a mild heat treatment at 60°C for 10 min to kill the germinated spores (Van Opstal et al., 2004). HHP treatment causes considerable damage to the germinated cells, making the germinated cells more sensitive to adverse environmental conditions such as heat (Smelt, 1998). However, in order to achieve a considerable reduction of bacterial spores by a pressure $\leq$ 600 MPa, heat treatment at temperatures $\geq$75–80°C should be applied as a subsequent step to HHP treatment (Smelt et al., 2002). Unfortunately, this two-step treatment is likely to cause coagulation of the milk, which is a common phenomenon when high levels of HHP treatment ($>$ 500 MPa) are used (Van Opstal et al., 2004).

Nonetheless, some pressure-resistant non-sporeforming bacteria, including certain strains of *Escherichia coli* and *Listeria monocytogenes*, require pasteurization at $>$ 600 MPa at $\geq$ 50°C to achieve a 6 log units reduction. Therefore, to meet the requirement for pasteurization of a 6 log unit reduction of microorganisms by treatments combining HHP and heat, either the HHP treatment step or heat treatment step should be increased accordingly above these specified levels (Van Opstal et al., 2004).

It is hypothesized that applying combinations of suboptimal factors that inhibit microbial growth in food would enhance food safety and stability while reducing any detrimental effects on the sensory and the nutritional qualities of the food product. HHP treatment could be used along with other preservation factors such as the application of mild heat or the addition of food preservatives. Some researchers have noticed that the inhibitory effect of HHP treatment (400 MPa) against *B. coagulans* spores was improved when applied in combination with mild heat at low pH and in the presence of nisin (Roberts and Hoover, 1996). Nisin exhibits antibacterial activity by creating pores in the bacterial cell membrane. It is possible that HHP treatment intensifies the antimicrobial activity of nisin by improving the permeability of the inner cell membrane, and permitting diffusion of the compound into the cell.

The effectiveness of HHP for inactivation of other *Bacillus* spp. can be enhanced with a combined heat or antimicrobial treatment. For example, *B. cereus* spores can be inactivated with a milder HHP treatment if mild heating and antimicrobial agents are employed (Shearer et al., 2000). Approximately a 5 log units reduction in *B. cereus* spores out of an initial spore count of about 6.5 log units was achieved by HHP treatment at 392 MPa when combined with 0.1% sucrose laurate and heating at 45°C for 10 min.

Improving the effectiveness of an HHP treatment by adding antimicrobial agents is more variable in the case of spore inactivation compared to vegetative cells. Both nisin and lysozyme can more readily permeate the cells of either Gram-positive or Gram-negative bacteria during or following HHP treatment because HHP inflicts sublethal injury. However, the same effect is not necessarily observed with the treatment of spores. For example, the use of antimicrobial treatment (nisin or lysozyme) in combination with HHP (300 or 400 MPa) did not improve HHP treatment efficiency against *B. cereus* spores. However, when a germinative cycle (60 MPa/120 min) was applied the HHP inactivation of *B. cereus* spores increased markedly (Figure 3.4).

This synergistic interaction between HHP treatment and nisin was observed in a model cheese system. It was suggested that the HHP germinative cycle may have triggered increased spore coat permeability, which facilitated nisin access to the cytoplasmic membrane. Nonetheless, the use of the additive lysozyme in cheese in combination with

HHP treatment of HHP germinated cells did not result in a synergistic inhibition of *B. cereus* spores (López-Pedemonte et al., 2003). The inhibitory effect of the HHP-nisin combination on *B. cereus* persisted for a 15-day storage period at 8°C, whereas the effect of HHP-lysozyme combination did not. Considering the psychrotropic nature of some *B. cereus* strains and their extraordinary heat resistance, *B. cereus* could cause food poisoning despite pasteurization and refrigerated storage, if complete inactivation of the spores is not ensured. Therefore, HHP treatment in combination with nisin offers a possible solution to the production of mild cheese and milk products that are not acidified to a satisfactory degree of safety and reasonable shelf life.

HHP has also been combined with pulsed electric field (PEF) treatments as a new approach to food preservation (Barbosa-Cánovas et al., 2000). The combination of HHP treatment (200 MPa at 40°C for 24 hr) with PEF treatment of 25 Kv/cm and supercritical fluid treatment resulted in $>$ 3 log units inactivation of *B. cereus* in an inoculated glycerol solution. In comparison, vegetative microorganisms such as *E. coli* and *Staphylococcus aureus* were not recoverable following similar combination treatments (Spilimbergo et al., 2003). PEF destabilizes bacterial cell membranes by a phenomenon known as electroporation (Barbosa-Cánovas et al., 2000). Electroporation is more difficult with spores than with vegetative bacterial cells because the spore cortex is fairly resistant. With electroporation, vegetative bacterial cells become more vulnerable to other preservation stresses such as HHP treatment and supercritical CO_2 technique. As expected, high-intensity PEF only results in a mild shocking of the *B. cereus* spore cortex and limited electroporation. Thus the subsequent effect of HHP treatment and supercritical CO_2 elicited incomplete inactivation of *B. cereus* spores (Spilimbergo et al., 2003).

Other combined treatments effective with vegetative cells have been attempted in spore suspensions. For example, vegetative bacterial cells can be inactivated at moderate temperature and pressure when CO_2 treatment is used, whereas pressure treatment alone has a little effect (Knorr and Heinz, 2001). The inactivation of different *Bacillus* spp. by pressure, heat, and CO_2 treatment was studied (Watanabe et al., 2003). *B. cereus* spores have been shown to be more sensitive to a combination of CO_2 treatment and moderate HHP (30 MPa) at 35°C compared to other *Bacillus* spp. such as *Geobacillus stearothermophilus*, *B. coagulans* and *B. subtilis* (Table 3.2). *G. stearothermophilus* was found to be the most

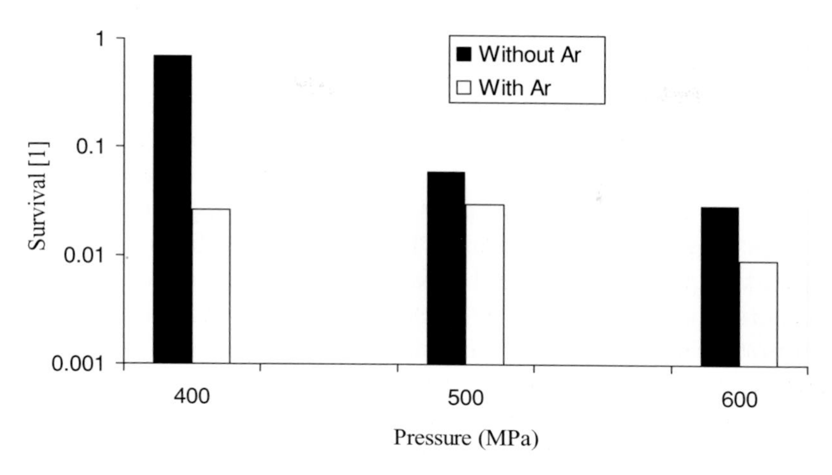

Figure 3.5. Effect of added argon (Ar) with high pressure on inactivation of *Bacillus cereus* at a treatment temperature of 30°C. (Adapted from Fujii et al., 2002.)

heat resistant compared to other *Bacillus* spp. However, *B. subtilis* was the most resistant to the combination of CO_2 treatment with moderate pressure (30 MPa) at 35°C compared to the other *Bacillus* spp evaluated (Watanabe et al., 2003). The extent of inactivation of *Bacillus* spp. spores by combined HHP-CO_2 treatments can be enhanced by increasing the pressurization temperature. CO_2 in a combination treatment decreases the processing temperature and pressure needed to inactivate bacterial spores. This is because CO_2 decreases the inactivation energy (kJ/mol) needed for spore inactivation.

The combined effect of water ordering with high pressure on the inactivation of *B. cereus* spores has also been studied (Fujii et al., 2002). Water ordering is a phenomenon whereby the crystallinity of the liquid structure of water is increased. Water ordering can be triggered by the addition of a rare gas such as argon. The addition of argon to bacterial cultures containing *B. cereus* spores was reported to enhance spore inactivation at relatively low pressure treatments (400 MPa) and at relatively low temperatures (20–30°C; see Figure 3.5 and Fujii et al., 2002). It was hypothesized that the inactivation of *B. cereus* spores depends on a balance between two phenomena; first, hydrophobic hydration that occurs at relatively low pressures (100–150 MPa) that impairs bacterial spore pressure resistance, and second, protein denaturation that occurs more intensively at high pressures ($>$500 MPa).

Summary

Several factors influence the effectiveness of HHP for the inactivation of bacterial spores. HHP is a relatively new food processing technology that has the potential to inactivate several important foodborne pathogens, mostly vegetative cells, and sporeforming microorganisms such as *B. cereus*. HHP treatment has the potential for use in the food industry to produce value-added food products with enhanced sensory characteristics and quality attributes that might not be attainable through conventional thermal processing or other treatment methods. The nutritional content of pressure-treated foods is generally higher than thermally treated foods.

Because of the risk of survival and possible growth of spores or of hazardous food pathogens (that might include toxin production), there have been few applications to date for the high pressure treatment of low-acid foods. Of particular concern in low-acid foods are *B. cereus* and *Clostridium botulinum* spores, because of their putative resistance to heat. By constructing processes that combine high pressure and heat, it may be possible to sterilize low-acid foods using pressure ranges that are commercially feasible in place of a high heat process that often results in undesirable changes in foods that degrade sensory quality attributes and the nutritional value of food products. Sublethal injury is an important consideration for any preservation method. Given favorable conditions, such as prolonged storage in a suitable, nutrient-rich substrate, sublethally injured cells may be able to recover. This recovery is of a particular significance for a psychrotropic pathogen such as *B. cereus* that can survive and grow under refrigeration conditions and pose risks to consumers.

B. cereus spores are sensitive to anti-microbial treatment such as nisin, especially when used in combination with HHP treatment, possibly because HHP facilitates the penetration of nisin to the inner cell membrane. Applying oscillatory pressure treatments between low and high pressures could reduce the resistance of *B. cereus* spores considerably.

Acidity greatly influences the extent of *B. cereus* inactivation by HHP treatment. *B. cereus* spores are more resistant to the direct effect of HHP treatment at neutral pH values. However, an acidic pH of the medium may be beneficial by inhibiting the recovery of sublethally injured cells.

In foods, the effect of intrinsic factors such as food composition (moisture content, fat, salt), water activity, pH, and acidity determine microbiological safety and stability of food during and after HHP treatment. Also, it is worthwhile to mention that results of HHP inactivation of *B. cereus* spores in a buffer or media cannot be extrapolated to real foods.

A problem that stands unresolved is the existence of a considerable fraction of recalcitrant super-dormant spores that remains ungerminated even after a prolonged pressurization. Finding ways to induce *B. cereus* spore germination, especially the super-dormant fraction, will be necessary if we are to obtain a reasonable understanding of the physiological response of *B. cereus* when exposed to high pressure. A prudent way to facilitate HHP-induced spore inactivation may be to induce germination under one pressure regime, and then to inactivate the germinated spores under another. The ability to achieve a high level of *B. cereus* spore germination in pressure-treated milk products supports the idea that it may be possible to achieve a 6 log reduction of these spores by conducting a treatment subsequent to the germination step, one that kills the germinated spores. Applying a pulsed or alternating pressure treatment has demonstrated a significant reduction in *B. cereus* spores compared to treatments at a constant pressure. Research to date on HHP inactivation of *B. cereus* and other sporeforming spoilage and pathogenic bacteria is rare, especially in real food products. Therefore, more research is still needed before the suitability of using HHP for preparing processed foods can be confirmed.

References

Agata, N., M. Ohta, M. Mori, and K. Shibayama. 1999. Growth conditions of an emetic toxin production by *Bacillus cereus* in a defined medium with amino acids. *Microbiology and Immunology* 43:15–18.

Barbosa-Cánovas, G.V., M.D. Pierson, Q.H. Zhang, and D.W. Schaffner. 2000. Pulse electric fields. *Journal of Food Science* supplemental:65–81.

Black, E.P., K. Koziol-Dube, D. Guan, J. Wei, B. Setlow, D.E. Cortezzo, D.G. Hoover, and P. Setlow. 2005. Factors influencing germination of *Bacillus subtilis* spores via activation of nutrient receptors by high pressure. *Applied and Environmental Microbiology* 71:5879–5887.

Choma, C., T. Clavel, H. Dominguez, N. Razafindramboa, H. Soumille, C. Nguyen-the, and P. Schmitt. 2000. Effect of temperature on growth characteristics of *Bacillus cereus* TZ415. *International Journal of Food Microbiology* 55:73–77.

Cléry-Barraud, C., A. Gaubert, P. Masson, and D. Vidal. 2004. Combined effects of high hydrostatic pressure and temperature for inactivation of *Bacillus anthracis* spores. *Applied and Environmental Microbiology* 70:635–637.

Fujii, K., A. Ohtani, J. Watanabe, H. Ohgoshi, T. Fujii, and K. Honma. 2002. High pressure inactivation of *Bacillus cereus* spores in the presence of argon. *International Journal of Food Microbiology* 72:239–242.

Granum, E., and T. Lund. 1997. *Bacillus cereus* and its food poisoning toxins. *FEMS Microbiology Letters* 157:223–228.

Hayakawa, K., T. Kanno, M. Tomito, and Y. Fujio. 1994. Oscillatory compared with continuous high pressure sterilization. *Journal of Food Science* 69:164–167.

Hoover, D.G., C. Metrick, A.M. Papineau, D.F. Farkas, and D. Knorr. 1989. Biological effects of high hydrostatic pressure on food microorganisms. *Food Technology* 43:99–107.

Jay, J.M. 2000. *Modern Food Microbiology*. Gaithersburg, MD: Aspen Publishers Inc.

Johnson, K.M. 1984. *Bacillus cereus* food-borne illness. *Journal of Food Protection* 47:145–153.

Katsui, N., T. Tsuchido, M. Takano, and I. Shibasaki. 1981. Effect of pre-incubation temperature on the heat resistance of *Escherichia coli* having different fatty acid compositions. *Journal of General Microbiology* 122:357–361.

Knorr, D. 1995. "Hydrostatic pressure treatment of food: Microbiology." In: *New Methods of Food Preservation,* ed. G.W. Gould, pp. 157–175. London: Blackie Academic and Professional.

Knorr, D., and V. Heinz. 2001. "Development of nonthermal methods for microbial control." In: *Disinfection, Sterilization, and Preservation,* ed. S.S. Block, pp. 853–877. London: Lea and Febiger.

Leistner, L. 1992. Food preservation by combined methods. *Food Research International* 25:151–158.

López-Pedemonte, T., A. Roig-Sagues, A. Trujillo, M. Capellas, and B. Guamis. 2003. Inactivation of spores of *Bacillus cereus* in cheese by high hydrostatic pressure with the addition of nisin or lysozyme. *Journal of Dairy Science* 86:3075–3081.

Margosch, D., M.A. Ehrmann, M.G. Ganzle, and R.F. Vogel. 2004. Comparison of pressure and heat resistance of *Clostridium botulinum* and other endospores in mashed carrots. *Journal of Food Protection* 67:2530–2537.

Martens, B., and D. Knorr. 1992. Development of nonthermal processes for food preservation. *Food Technology* 46:124–133.

McClements, J.M., M.F. Patterson, and M. Linton. 2001. The effect of growth stage and growth temperature on high hydrostatic pressure inactivation of some psychotropic bacteria in milk. *Journal of Food Protection* 64(4):514–522.

Nakayama, A., Y. Yano, S. Kobayashi, M. Ishikawa, and K. Sakai. 1996. Comparison of pressure resistance of spores of six *Bacillus* strains with their heat resistances. *Applied and Environmental Microbiology* 62:3897–3900.

Oh, S., and M-J. Moon. 2003. Inactivation of *Bacillus cereus* spores by high hydrostatic pressure at different temperatures. *Journal of Food Protection* 66:599–603.

Paidhungat, M., B. Setlow, B. Daniels, D. Hoover, E. Papafragkou, and P. Setlow. 2002. Mechanisms of induction of germination of *Bacillus subtilis* spores by high pressure. *Applied and Environmental Microbiology* 68:3172–3175.

Patterson, M.F., M. Quinn, R. Simpson, and A. Gilmour. 1995. Sensitivity of vegetative pathogens to high hydrostatic pressure treatments in phosphate-buffered saline and foods. *Journal of Food Protection* 58(5):524–529.

Prestamo, G., M. Lesmes, L. Otero, and G. Arroyo. 2000. Soybean vegetable protein (tofu) preserved with high pressure. *Journal of Agriculture and Food Chemistry* 48:2943–2947.

Raso, J., G. Barbosa-Cánovas, and B.G. Swanson. 1998a. Sporulation temperature affects initiation of germination and inactivation by high hydrostatic pressure of *Bacillus cereus*. *Journal of Applied Microbiology* 85:17–24.

Raso, J., M.M. Gogora-Nieto, G. Barbosa-Cánovas, and B.G. Swanson. 1998b. Influence of several environmental factors on the initiation of germination and inactivation of *Bacillus cereus* by high hydrostatic pressure. *International Journal of Food Microbiology* 44:125–132.

Roberts, C.M., and D.G. Hoover. 1996. Sensitivity of *Bacillus coagulans* spores to high hydrostatic pressure, heat, acidity and nisin. *Journal of Applied Bacteriology* 81:363–368.

Setlow, B., C.A. Loshon, P.C. Genest, A.E. Cowan, C. Setlow, and P. Setlow. 2002. Mechanisms of killing spores of *Bacillus subtilis* by acid, alkali and ethanol. *Journal of Applied Microbiology* 92:362–375.

Setlow, P. 1994. Mechanisms which contribute to the long-term survival of spores of *Bacillus* species. *Journal of Applied Bacteriology* 76 (supplemental):49S–60S.

Shearer, A.E., C.P. Dunne, A. Sikes, and D.G. Hoover. 2000. Bacterial spore inhibition and inactivation in foods by pressure, chemical preservatives, and mild heat. *Journal of Food Protection* 63:1503–1510.

Shigehisa, T., T. Ohmori, A. Saito, S. Taji, and R. Hayashi. 1991. Effects of high hydrostatic pressure on characteristics of pork slurries and inactivation of microorganisms associated with meat and meat products. *International Journal of Food Microbiology* 12:207–215.

Shinagawa, K., J. Sugiyama, T. Terada, N. Matsusaka, and S. Sugii. 1991. Improved methods for purification of enterotoxin produced by *Bacillus cereus*. *FEMS Microbiology Letters* 80:1–5.

Smelt, J.P.P.M. 1998. Recent advances in the microbiology of high pressure processing. *Trends in Food Science and Technology* 9:152–158.

Smelt, J.P.P.M., J.C. Hellemons, P.C. Wouters, and S.J.C. Gerwen. 2002. Physiological and mathematical aspects in setting criteria for decontamination of foods by physical means. *International Journal of Food Microbiology* 78:57–77.

Sojka, B., and H. Ludwig. 1994. Pressure induced germination and inactivation of *Bacillus subtilis* spores. *Die Pharmazeutische Industrie* 56(7):660–663.

———. 1997. Effects of rapid pressure changes on the inactivation of *Bacillus subtilis* spores. *Die Pharmazeutische Industrie* 59(5):436–438.

Spilimbergo, S., F. Dehghani, A. Bertucco, and N.R. Foster. 2003. Inactivation of bacteria and spores by pulse electric field and high pressure CO_2 at low temperature. *Biotechnology and Bioengineering* 82:118–125.

Van Opstal, I., C.F. Bagamboula, C.M. Suzy, S.C.M. Vanmuysen, E.Y. Wuytack, and C.W. Michiels. 2004. Inactivation of *Bacillus cereus* spores in milk by mild pressure and heat treatments. *International Journal of Food Microbiology* 92:227–234.

Watanabe, T., S. Furukawa, J. Hirata, T. Koyama, H. Ogihara, and M. Yamasaki. 2003. Inactivation of *Geobacillus stearothermophilus* spores by high-pressure carbon dioxide treatment. *Applied and Environmental Microbiology* 69:7124–7129.

Wuytack, E.Y., S. Boven, and C.W. Michiels. 1998. Comparative study of pressure-induced germination of *Bacillus subtilis* spores at low and high pressures. *Applied and Environmental Microbiology* 64:3220–3224.

Yuste, J., M. Capellas, D.Y. Fung, and M. Mor-Mur. 2004. Inactivation and sublethal injury of foodborne pathogens by high pressure processing: Evaluation with conventional media and thin agar layer method. *Food Research International* 37:861–866.

Chapter 4

Inactivation of *Bacillus* Spores at Low pH and in Milk by High Pressure at Moderate Temperature

Isabelle Van Opstal, Abram Aertsen, and Chris W. Michiels

Introduction and Scope

Dormant bacterial endospores are very resistant to anti-microbial agents and treatments such as high temperature, radiation, chemicals, and high pressure. Generally, the vigorous treatments that are required to destroy spores in foods at the same time also have detrimental effects on the nutritional and sensory properties of the food products. During the last few decades, food processors and scientists have worked on the development of novel food preservation processes that are less destructive to the flavor, texture, and nutritional qualities of food, yet ensure equal or better food stability and microbiological safety than conventional processes such as thermal pasteurization and sterilization. High pressure processing is among the most advanced of these techniques, and it has already been adopted at the commercial scale for specific applications. A weak point of high pressure as a food preservation treatment is that alone it does not kill spores efficiently, unless it is combined with elevated temperatures ($> 70°C$). However, this combination will partly offset the advantages of the process with regard to the preservation of quality attributes. Therefore, it remains of interest to investigate why spore inactivation by high pressure at moderate temperatures ($< 60°C$) is inefficient, and to attempt to improve this efficiency.

It has been well established that spore inactivation at high pressures and moderate temperatures proceeds as a two-step process, in which spores are first induced to germinate by high pressure, and then the germinated spores, due to their increased overall stress sensitivity, are subsequently killed by high pressure (Clouston and Wills, 1969; Sale et al., 1970). The mechanisms underlying pressure-mediated spore germination are already understood in some detail (Wuytack et al., 1998, 2000; Setlow, 2003). Similar to any other method of inducing spore germination, high pressure is unable to germinate a certain fraction of so-called super-dormant spores, and this is the major reason for the inefficient spore inactivation.

In this chapter, we will focus on some of our work on the efficiency of high pressure at moderate temperature to induce germination and inactivation of *Bacillus* spores in acidic conditions on the one hand, and in milk as a model low-acid food on the other hand. In the first part, we will also describe the use of recombinantly expressed green fluorescent protein as a reporter of spore physiology during high pressure treatment. This approach will provide insights in the changes that take place in the spore core during high pressure treatment and/or acid challenge of spores and help us understand the causes of spore inactivation. In the second part, we address the possibility that the efficiency of spore germination would be enhanced by combining high pressure treatment with chemical germinants. This will be studied in a buffer system first, and subsequently in milk, in which we assume that germinants are naturally present.

Use of Green Fluorescent Protein (Gfp) as a Probe for Studying the Fate of Spores under High Hydrostatic Pressure

Spore germination is accompanied by dramatic changes in the physiochemical environment of the spore core, such as a decrease in ionic strength and an increase in pH and hydration. These changes are expected to also affect the fluorescence properties of Gfp inside the spore, and therefore we anticipated that it should be possible to use Gfp present in recombinant spores as a fluorescent probe of the spore core environment, for example, to monitor spore germination or to monitor acidification of the spore core during acid challenge. We investigated this idea in a number of experiments with Gfp containing spores, first at neutral pH, and subsequently at low pH. The strain used was *B. subtilis* CW335, carrying the *gfp* gene under control of an artificial promoter (sspE-2G) that supports a high level of transcription during sporulation, but that is silent in vegetative cells (Webb et al., 1995). As a result, spores of strain CW335 contain Gfp and exhibit green fluorescence upon excitation with UV light, but vegetative cells do not.

Observations at Neutral pH

Changes in Gfp Fluorescence upon Germination

First, we determined the effect of spore germination on Gfp fluorescence. In ungerminated spores of *B. subtilis* CW335, Gfp exhibited an excitation peak at 390 nm, similar to that occurring in free solution. The minor excitation peak of Gfp in free solution at 470 nm (Cubitt et al., 1995) was, however, not observed in the spores (Figure 4.1A). When excited at 390 nm, Gfp in the spores showed an emission maximum at 509 nm and a shoulder at 540 nm (Figure 4.1B). Moreover, it can be seen on this spectrum that Gfp fluorescence of the spore suspensions increased when spores were germinated by high pressure.

There are, however, some problems if one would like to use this fluorescence increase to monitor spore germination. Firstly, the useful dynamic range is limited to approximately 1 decade. Once more than 90% of the spores are germinated, it becomes difficult to assess a further increase accurately. This can be clearly seen in Figure 4.2. The signal increase for the 2nd decade of germination will indeed be ten times lower than that for the 1st decade. Secondly, the rehydration of the core, which takes place during germination and which is probably the cause for the increased fluorescence, presumably makes Gfp more susceptible to pressure denaturation and thereby may interfere with the assay, particularly when using high pressures. The increased intrinsic fluorescence of Gfp due to rehydration of the spore core may then be offset by partial denaturation of Gfp.

Changes in Heat Stability of Gfp upon Germination

The small increase in Gfp fluorescence that accompanies spore germination is not a suitable parameter for quantitative measurement of germination. However, we anticipated that the hydration of the spore protoplast during germination would strongly increase the heat sensitivity of Gfp compared to that in ungerminated spores. It should then

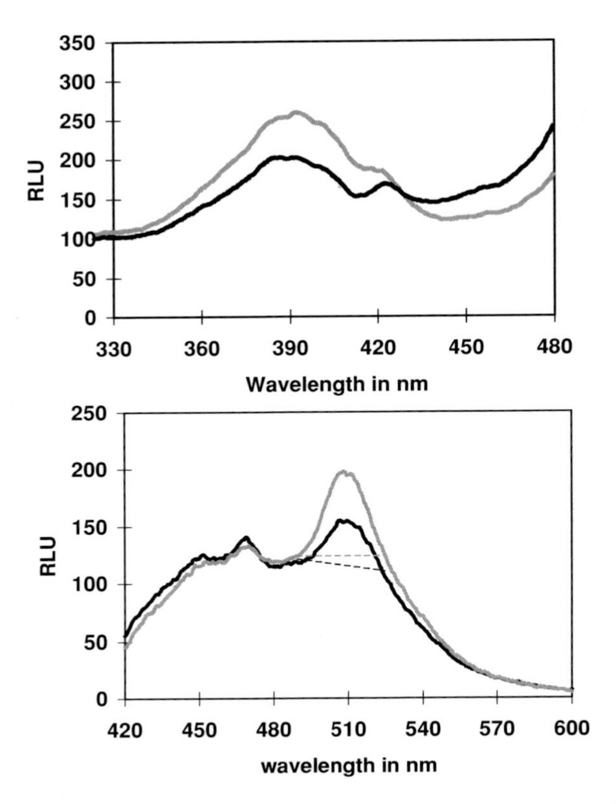

Figure 4.1. Excitation spectrum with emission at 509 nm (A) and emission spectrum with excitation at 390 nm (B) of ungerminated (dark line) and pressure germinated (grey line) *B. subtilis* CW335 spore suspensions (RLU: relative light units; dashed line: baseline). Pressure treatment was at 100 MPa and 40°C.

be possible to selectively inactivate Gfp in the germinated spores while leaving Gfp in the ungerminated spores intact, and to determine spore germination quantitatively by measuring the residual fluorescence of a germinated spore suspension after a heat treatment.

In Figure 4.3, the heat stability of Gfp in pregerminated and in ungerminated spores is compared. When 15-min heat treatments are applied, Gfp is stable (fluorescence retention $> 80\%$) in ungerminated spores at least up to 80°C. In pregerminated spores, in contrast, fluorescence is completely eliminated (fluorescence retention $< 2\%$) at 80°C.

The increased heat sensitivity of Gfp upon pressure-induced germination was also confirmed by epifluorescence and phase-contrast

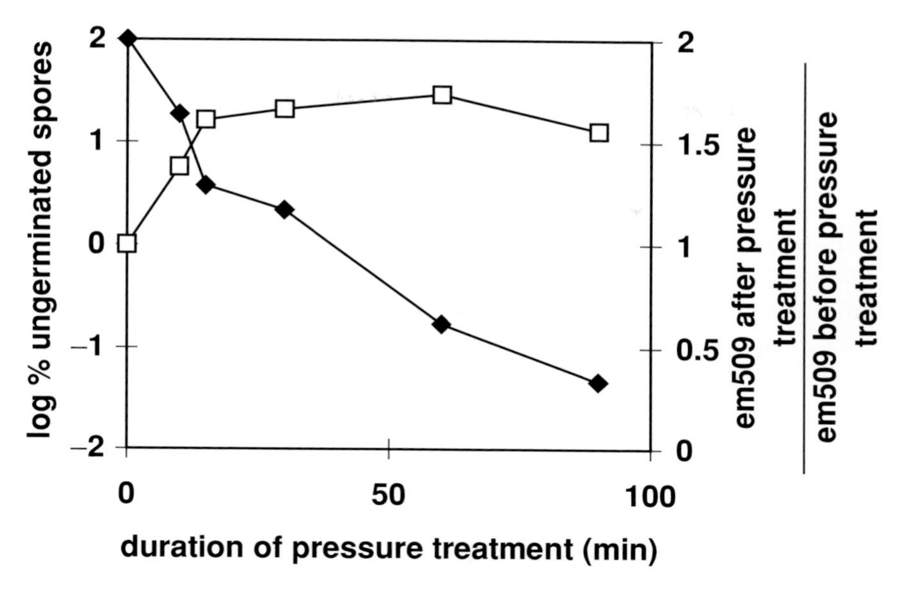

Figure 4.2. Degree of germination (♦) and Gfp fluorescence (□) of *B. subtilis* CW335 spore suspensions. Pressure treatment was at 100 MPa and 40°C.

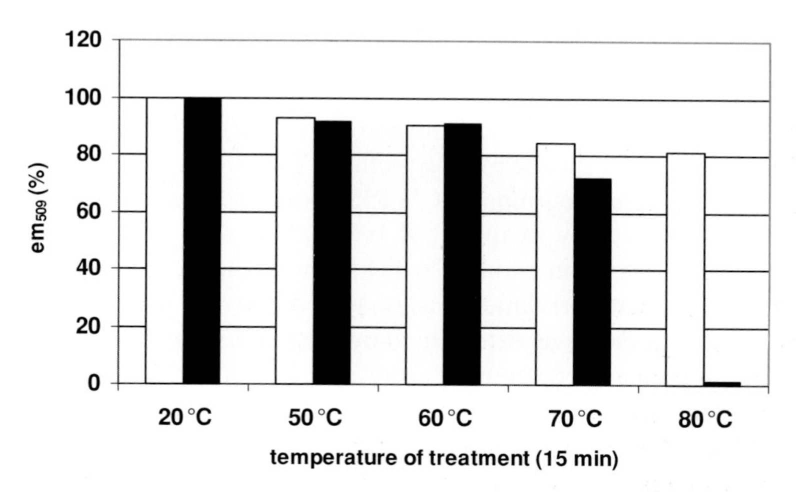

Figure 4.3. Residual fluorescence of Gfp after heat treatment of ungerminated (□) and pressure pregerminated (■) *B. subtilis* CW335 spore suspensions. Pregerminated spores were obtained by pressure treatment at 100 MPa and 40°C for 30 min. em_{509} at 20°C for ungerminated and pregerminated spores was set at 100%.

microscopy. After pressure treatment, germinated spores became phase-dark but retained Gfp fluorescence. However, after an additional heat treatment (80°C/15 min), the phase-dark spores lost their fluorescence and only phase-bright spores retained fluorescence (Wuytack and Michiels, 2001).

Making use of the increased heat sensitivity of Gfp upon pressure-induced germination, we then attempted to develop a novel rapid method for determining the degree of germination in a suspension of Gfp containing spores after high pressure treatment. Briefly, the pressurized spore suspensions were first subjected to a heat treatment (80°C/15 min) by immersion in a water bath to eliminate the Gfp fluorescence in the germinated spores. Only the ungerminated spores will remain fluorescent after this treatment. Then, the emission spectrum (excitation at 390 nm) of pressurized and unpressurized samples was recorded. The peak heights at 509 nm (em_{509}) were determined relative to a baseline drawn from 490 nm to 525 nm (Figure 4.1B). The percentage of ungerminated spores was then expressed as:

$$\%\ ungerminated\ spores = \frac{em_{509}\ after\ pressure\ treatment}{em_{509}\ before\ pressure\ treatment} * 100\% \qquad (4.1)$$

The accuracy and sensitivity of the Gfp method for measuring spore germination was determined by comparison with the two standard methods: the OD_{600} method and the plating method. To this end, the degree of germination of twenty *B. subtilis* CW335 spore suspensions, germinated to various degrees by treatment at 100, 225, or 400 MPa for different times, was assessed in parallel by the three methods. In a full logarithmic plot, a linear correlation was observed between the percentage of ungerminated spores as determined by the plating method and the percentage ungerminated spores as determined by both the OD_{600} and Gfp method (Figure 4.4). However, at high levels of germination, deviations occur because the signal in both optical measurements approaches the noise level. As can be seen in Figure 4.4 (data points in grey), the OD method will lead to an overestimation of germination at high levels of germination, while the Gfp method will result in an underestimation, compared to the plating method.

While making an exemption for the deviations occurring in the data points corresponding to high levels of germination, a good correlation

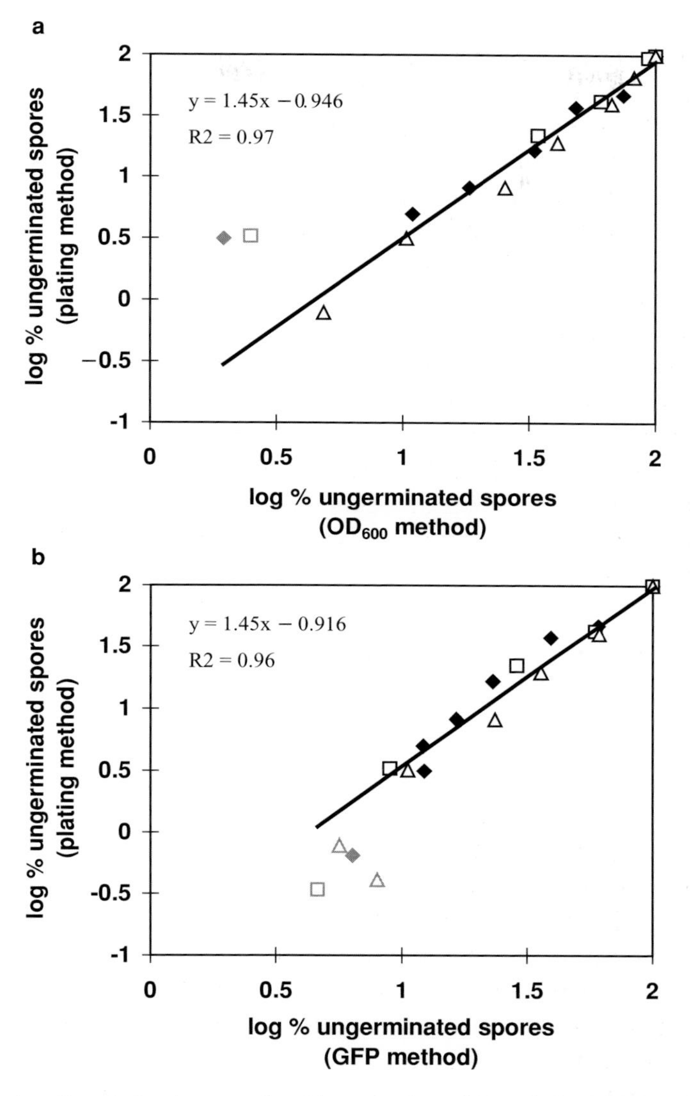

Figure 4.4. Correlation between log % ungerminated *B. subtilis* CW335 spores determined with the plating method and the OD_{600} method (A) or the Gfp method (B). The spore suspensions were germinated at 100 MPa (♦), 225 MPa (□), or 400 MPa (Δ). Grey data points indicate outliers and were not taken into account in generating the trendline.

was established for both the OD_{600} method and the Gfp method ($R2 = 0.97$ and 0.96, respectively) when compared to the plating method, thus testifying to the relative accuracy of these rapid optical methods. A limitation of both optical methods, however, is their limited dynamic range (approximately 2 log units), which is due to the fact that when more than 99% of the spores in the suspension are germinated, the signal of both optical methods falls below the noise level. However, we anticipate that the sensitivity of the Gfp method may be increased at least 10-fold by the use of new modified Gfps with strongly increased fluorescence.

In this respect, Crameri et al. (1996) constructed some new modified Gfps with increased fluorescence by improving the codon usage and performing recursive cycles of DNA shuffling followed by screening for the brightest *Escherichia coli* colonies. Due to the changes in the amino acid sequence, most of the mutant protein is soluble and active in *E. coli*, whereas most of the wild-type Gfp ends up in inclusion bodies, unable to activate its chromophore. The emission and excitation maxima of the modified Gfp were the same as those of the wild-type Gfp. In addition, despite their obvious biological utility, the spectral properties of wild-type Gfp are far from ideal, and numerous Gfp-mutants with different excitation and emission spectra have been developed. For example, new modified Gfps have been developed with spectral properties to be more similar to that of fluorescein isothiocyanate (FITC), which is well detected by modern microscopic filter sets (Heim et al., 1995). In this respect, red-shifted Gfp mutants are significantly brighter when excited with blue light compared to wild-type Gfp and have improved properties for growth at higher temperatures (Heim et al., 1995). Mutant Gfps with different excitation and emission spectra, resulting in an extensive pallet of colors ranging from blue to yellow, open the prospects for double and triple labeling of cells. However, while these modified Gfps have proved useful in studies with relatively large eukaryotic cells, many of them are currently not suitable for use in microorganisms, because they are not bright enough.

Currently, some attempts have, however, been made to resolve this problem. For example, Lewis and Marston (1999) introduced a series of plasmid vectors for the construction of Gfp fusions that integrate by single or double cross-over into the chromosome of *B. subtilis*. Because of the quite specific excitation and emission wavelengths characteristic of Gfp fluorescence, the use of Gfp for measuring germination in

B. subtilis spores should allow the analysis of spore germination in mixed bacterial suspensions and in slightly turbid media-like foods. Furthermore, techniques such as fluorescence microscopy based on the fluorescence of Gfp and combined with image analysis, fluorescence recovery after photobleaching (FRAP) of Gfp, or flow cytometry using Gfp have been recently developed that allow the assessment of Gfp fluorescence of single spores or bacteria within a large population. Moir (2003), for example, used FRAP of Gfp to report on protein mobility in single spores. Lun and Willson (2004) investigated the interaction between type 2 *Streptococcus suis* and swine phagocytes during infection of the natural host by using Gfp as a specific marker to observe the challenge organism. Moreover, these techniques, together with the potentially interesting use of Gfp for the measurement of spore germination, as reported in this section, may also open perspectives to the assessment of germination at the individual cell (or spore) level. Furthermore, the kinetics of spore inactivation will probably be more comparable to the kinetics of inactivation of Gfp inside the spore protoplast than of an enzyme in free solution. Finally, variants of the Gfp and Gfp-like proteins, such as red fluorescent proteins, have recently been isolated and are currently available for applications in multicolor labeling and fluorescence energy transfer (Terskikh et al., 2000).

Observations at Low pH

Stability of Gfp in Spores and Pregerminated Spores at Low pH

Gfp in recombinant *Bacillus* spores, as highlighted in the section "Observations at Neutral pH," is not only a useful tool for measuring germination in *B. subtilis* spores but also a potentially interesting fluorescent probe for measuring intracellular pH levels, because Gfp fluorescence in free solution was reported to be unstable at pH < 5.5 (Bokman and Ward, 1981). In a first set of experiments, we determined the effect of low pH on the stability of Gfp in ungerminated and pregerminated spore suspensions, and, at the same time, determined survival of the cells under these conditions. Fluorescence emission values (E at $\lambda_{ex} = 390$ nm, and $\lambda_{em} = 509$ nm) recorded at any given pH (pH $=$ x) were determined from the emission spectra as described above; then they were transformed into normalized emission values (E_x^n), which are defined as the ratio of the difference in E_x (the emission at pH $=$ x) and

E_3 (the emission at pH = 3) to the difference in E_7 (the emission value at pH = 7) and E_3, as follows:

$$E_x^n = \frac{E_x - E_3}{E_7 - E_3} \tag{4.2}$$

Intact ungerminated spores suspended in a pH 3.0 buffer remained fully fluorescent for at least 3 hr. In contrast, when spores were first germinated at pH 7.0 by pressure and subsequently resuspended at low pH (pH < 5.0) for 10 min, Gfp fluorescence was completely lost (Figure 4.5). Gfp fluorescence was even more sensitive in spores pregerminated at 600 MPa than in spores pregerminated at 100 MPa. Resuspension of the germinated spores in buffer pH 7.0, after exposure to pH 3, did not restore Gfp fluorescence, even after prolonged incubation at pH 7.0, suggesting irreversible damage to the Gfp protein (data not shown). The Gfp fluorescence in spores also disappeared when dormant spores were subjected to a pressure treatment at pH = 3 (data not shown).

Interestingly, Figure 4.5 also shows that, in general, loss of Gfp fluorescence correlates well with loss of viability in these experiments, in which cell suspensions were plated after 1 hr of incubation at the respective pH at room temperature. Irrespective of the germination method, all germinated spores were found to be sensitive to low pH, but not equally sensitive. The sensitivity increased in the order alanine-germinated spores, 100 MPa-germinated spores, and 600 MPa-germinated spores (Figure 4.5). The viability of ungerminated spores, as expected, was not affected by low pH challenge.

These observations agree with the postulate of Swerdlow et al. (1981) that there is a strong permeability barrier in dormant spores for movement of charged molecules that is breached early in spore germination. The loss of Gfp fluorescence in germinated spores exposed to acid suggests there is an influx of protons and acidification of the protoplast. Furthermore, our results show the 600 MPa germinated spores to be even more sensitive to this intracellular acidification than spores germinated at 100 MPa or by alanine. This, it turns out, is well explained by the earlier observation that the 600 MPa treatment, although triggering germination, inhibits the rapid onset of ATP production that normally takes place early after germination (Wuytack et al., 1998). As a result, spores germinated at 600 MPa will be deficient in pH homeostasis, because they lack the energy required to pump incoming protons out of the protoplast.

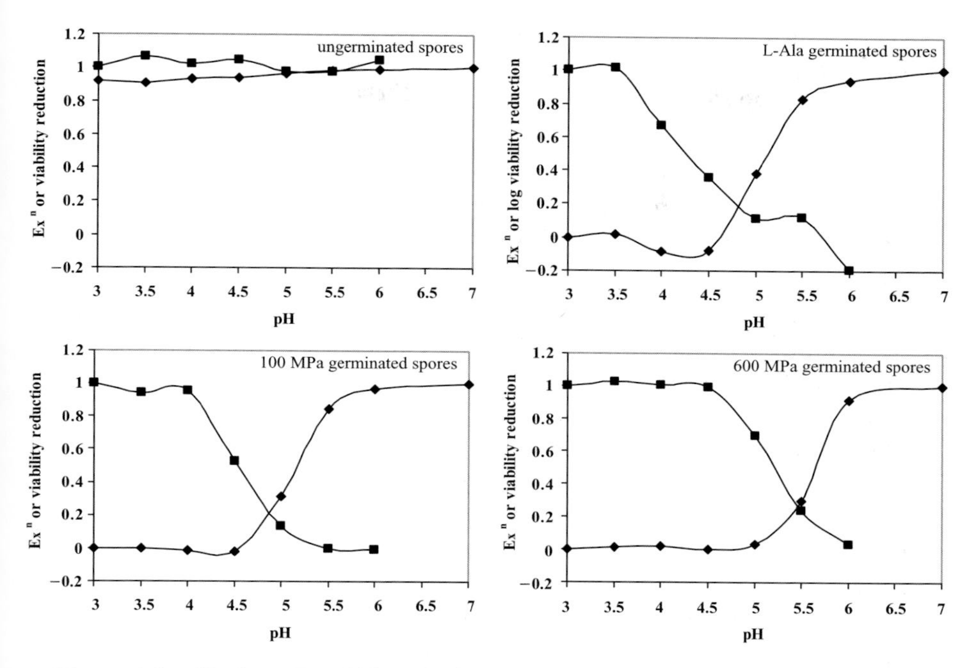

Figure 4.5. The low pH stability of Gfp in *B. subtilis* CW335 spores (♦) and the low pH sensitivity of *B. subtilis* CW335 spores after 1 hr of incubation at the indicated pH (■) that are ungerminated or pregerminated at pH 7.0 by alanine, 100 MPa, or 600 MPa. E^n_x: normalized fluorescence emission at pH = x; viability reduction: normalized as log (viability reduction at pH = x)/log (viability reduction at pH 3.0).

It should be noted, however, that not all of the 600 MPa germinated spores (log number of ungerminated spores = 3.78) were inactivated at low pH (log number of surviving spores = 5.53). For the alanine and 100 MPa germinated spores, in contrast, complete inactivation was achieved at pH 3.0. Thus, it seems that the physiology of spores germinated at 600 MPa displays a large heterogeneity that requires further investigation.

Stability of Gfp in Spores upon Heating at Low pH

In the above section, we investigated the use of Gfp as a probe to monitor changes in the intracellular environment of spores and pregerminated spores upon acid challenge at ambient temperature, and tried to correlate these changes to spore viability. Here, the same general approach was used to study spores subjected to the combined effect of heat and low pH. It is known that spores are more easily inactivated by heat at low

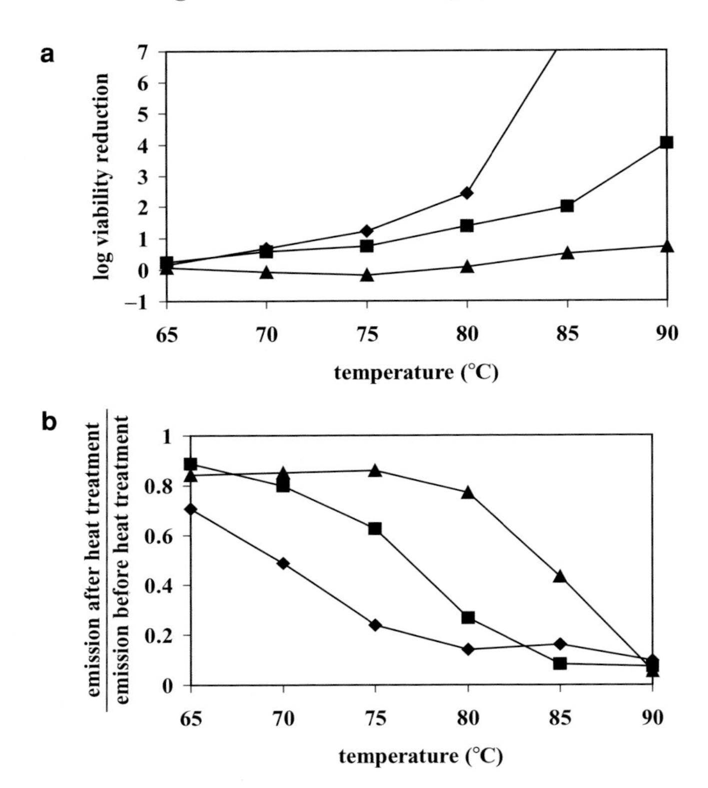

Figure 4.6. Viability loss (A) and loss of Gfp fluorescence (B) in ungerminated *B. subtilis* CW335 spores suspended in 50 mM HEPES buffer at pH 3.0 (♦), pH 4.5 (■), and pH 7 (▲) and subjected to different successive heat treatments of 10 min at 65, 70, 75, 80, 85, and 90°C. (Reprinted from Wuytack et al., 2001, with permission from Elsevier.)

than at neutral pH (Fernandez et al., 1994). Suspensions of dormant spores at pH 3.0, 4.5, and 7.0 were subjected to successive 10-min heat treatments at increasing temperatures. After each heat treatment, the reduction in viability (Figure 4.6A) and the fluorescence of the spore suspensions (Figure 4.6B) were determined.

It was confirmed that spores suspended at low pH are more sensitive to heat than those resuspended at neutral pH (Figure 4.6A). Furthermore, Gfp fluorescence was much more heat sensitive in spores at low pH than at neutral pH (Figure 4.6B). This Gfp quenching provides direct evidence that heating at low pH causes acidification of the spore protoplast, and suggests that this is the cause of the ensuing spore inactivation. As discussed in the previous section, acidification of the spore protoplast does not take place at ambient temperature. In general, loss of Gfp fluorescence in this experiment correlates well with the loss of viability occurring in different pH conditions and for temperatures up to 80°C. Above that temperature, Gfp is inactivated both in acid and neutral pH, while the spores remain viable at least up to 90°C at pH 7.0. These results show that the rapid fluorescence response to pH changes observed with Gfp in free solution (Elsliger et al., 1999) also occurs in *B. subtilis* spores, making Gfp a useful probe for monitoring intracellular pH in spores. This application adds to earlier proposed applications of Gfp as an intracellular pH probe (Kneen et al., 1998; Llopis et al., 1998; Hanson et al., 2002), and of Gfp and calmodulin chimaera as fluorescent indicators of cellular Ca^{2+} concentrations (Miyawaki et al., 1997; Romoser et al., 1997).

Stability of Spores upon High Pressure Treatment at Low pH

In this section we report on spore germination and inactivation by high pressure at low pH. These experiments were conducted with the non-Gfp-containing spores of *B. subtilis* strain LMG 7135. Spore suspensions in buffers at pH 3.0–8.0 were subjected to 100 or 600 MPa at 40°C for 20 min and either plated directly to enumerate survivors, or plated after an additional heat treatment to count the ungerminated survivors (Figure 4.7; see also Wuytack and Michiels, 2001).

Over the tested range of pH, the spores survived the treatments at 100 and 600 MPa and 40°C well, since the reduction in viability was never more than 5-fold (Figure 4.7). When germination was determined on the basis of the difference in heat sensitivity between ungerminated and germinated spores at 80°C (10 min), more germination was found to occur at low than at neutral pH. However, this finding was not in agreement with optical density measurements, which showed no decrease after pressure treatment at low pH (pH $\leq$ 4.0), indicating that pressure-induced germination is in fact inhibited at low pH (data not shown).

We also analyzed the heat sensitivity of the pressure-treated spore suspensions at 60°C (10 min), and these are also shown in Figure 4.7. Interestingly, spores pressurized in low pH media (pH $\leq$ 4.0) were not inactivated by subsequent heating at 60°C, as opposed to being inactivated at 80°C. Conversely, spores pressure treated at neutral pH became

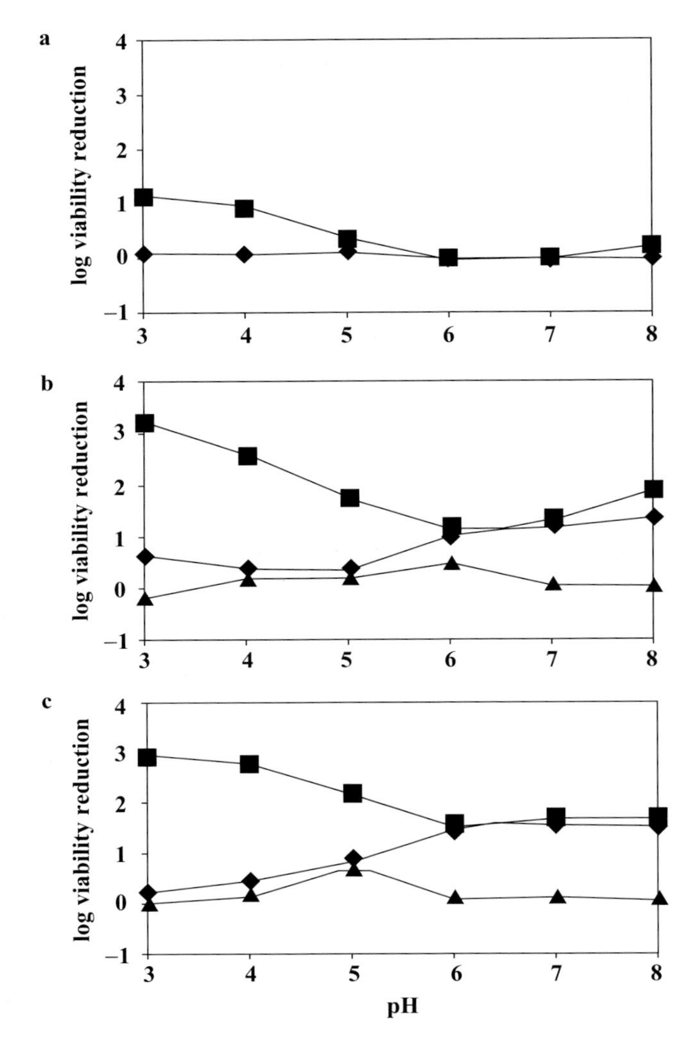

Figure 4.7. Inactivation of *B. subtilis* LMG7135 spores in buffers of different pH after a pressure treatment (20 min, 40°C) of 0.1 MPa (A), 100 MPa (B), or 600 MPa (C), followed by a heat treatment for 10 min at 20°C (control, ▲), 60°C (♦), or 80°C (■). Different buffers were used to cover the pH range: 100 mM HEPES (pH 3, 4, and 5), 50 mM NaH_2PO_4-citrate (pH 5, 6, and 7), and 50 mM potassium phosphate (pH 7 and 8). (Reprinted from Wuytack et al., 2001, with permission from Elsevier.)

sensitive to both 60°C and 80°C treatments. These findings led us to conclude that low pH inhibits high pressure induced spore germination, but renders spores more sensitive to heat by a mechanism different from germination. This induced heat sensitivity by high pressure at low pH is only of an intermediate level, however, since the spores remained resistant to heat treatment at 60°C, a temperature that rapidly kills truly germinated spores (Figure 4.7). Thus, low pH inhibits spore germination by high pressure, as has been previously observed for germination by chemical germinants with *B. subtilis, Bacillus cereus*, and *Bacillus megaterium* (Bender and Marquis, 1982).

As a possible explanation for this intermediate heat sensitivity of spores pressurized at low pH, we propose that pressure promotes the exchange of spore cations (Ca^{2+}, Mg^{2+}, K^+, Na^+, and Mn^{2+}) with extracellular protons, turning them into so-called H-spores. This cation exchange is also taking place at low pH and ambient pressures, but at a much slower rate. These H-spores have been reported to be more sensitive to high temperature (Bender and Marquis, 1982) and more resistant to pressure compared to native *B. subtilis* spores (Igura et al., 2003). When H-spores were prepared at ambient pressure, and then subjected to high pressure treatment at neutral pH, we found a reduced ability to germinate compared to normal spores (data not shown). This result may indicate that spore demineralization is the primary reason for the inhibition of spore germination at low pH, both under ambient conditions and under high pressure. In the normal germination process, release of Ca^{2+} ions from the spore protoplast is an essential step in the signaling cascade that leads to spore germination, because the Ca^{2+} ions are required for activation of spore cortex hydrolase. In accordance with this scheme, Igura et al. (2003) found that the pressure resistance of H-spores could be restored to the wild-type level upon remineralization with Ca^{2+} or Mg^{2+}, but not with Mn^{2+} or K^+.

The results in this section therefore suggest that pressure, just like elevated temperature (see previous section), promotes the formation of H-spores at low pH, presumably because the barrier properties of the spore membrane and cortex are disabled to some extent. This inference was also supported by experiments with the Gfp containing spores of *B. subtilis* CW335, which indicated a considerable decrease in the pH of the spore protoplast upon pressure treatment at low pH, although no pressure-induced germination takes place. However, our experiments do not exclude the possibility that the higher heat sensitivity of Gfp in spores exposed to low pH, compared to neutral pH, may result from partial rehydration of the spore core occurring at low pH, rather than from a low core pH itself. That spore core dehydration does indeed affect Gfp heat stability could already be concluded from the higher heat stability of Gfp in ungerminated than in germinated spores (Figure 4.3).

Conclusions

Gfp produced in recombinant spores can be used as an intracellular probe to investigate physicochemical changes in the spore core resulting from stress challenge. The decreased heat stability of Gfp that results from rehydration of the protoplast during spore germination can be used to monitor germination of spore suspensions and of individual spores. Furthermore, Gfp fluorescence can also be used to monitor acidification of the spore protoplast in low pH environments. Using this technique, we demonstrated that high pressure induced spore germination is inhibited under sufficiently acidic conditions, but that demineralization of the spore core is stimulated.

These findings have some implications for the high pressure preservation of acid foods. A pressure treatment conducted at low temperature ($\leq$ 40°C) will cause little or no direct inactivation of spores in these foods. However, a pressure pretreatment will make spores more sensitive to subsequent heating exposure, so that a considerably milder heat treatment can be applied to efficiently inactivate spores in acid food products. This two-step process is illustrated by the data in Figure 4.7 for *B. subtilis* spores suspended in buffers. A heat treatment of 80°C for 10 min caused 1 decade of inactivation in a spore suspension at pH 3, compared to 3 decades of inactivation when this spore suspension was pre-treated with a pressure of either 100 or 600 MPa at 40°C for 20 min. The pressure treatment in itself did not cause any inactivation.

For efficient spore killing in acid products, our results indicate that it would be even better to acidify a food product after the pressure treatment than before, if such an application is feasible, since pressure-induced germination, which sensitizes the spores, is inhibited at low pH. Acidification to pH = 3 will kill all spores germinated at 100 MPa, and a large fraction of the spores germinated at 600 MPa, even without the need for applying a heat treatment. Nevertheless, an important restriction remains that mild high pressure-temperature treatments followed by a low pH treatment produce only limited inactivation (3 to 6 decades), because of the limited level of germination that can be achieved under these circumstances.

Stimulation of High Pressure Induced Spore Germination and Inactivation by Germinants and in Milk

Effect of Heat Pretreatment and Germinants on High Pressure Induced Germination

In this section, we will focus on non-acid conditions again. High pressure treatment of non-acid foods at slightly to moderately elevated temperatures ($< 50^{\circ}C$) is generally considered to be a pasteurization treatment with the primary objectives of eliminating vegetative pathogens and reducing counts of spoilage organisms. However, if these products should require an extended refrigerated shelf life (e.g., $>$ 10 days), the issue of spore inactivation also becomes important, in order to prevent the outgrowth of psychotropic spore formers to high numbers. It has been documented that spore inactivation by high pressure at low and moderate temperatures proceeds via spore germination. In this section, we investigate whether germinants would stimulate germination of *Bacillus* spores by pressure. The motivation for this study stems from the finding that pressurization of *B. subtilis* triggers specific steps in the germinant-dependent germination pathways (Wuytack et al., 2000; Black et al., 2005). Relatively low pressures (up to about 300 MPa) have been shown to trigger the germinant receptors themselves, while higher pressures primarily act more downstream in the germination cascade by inducing the release of calcium dipicolinic acid (Ca^{2+}-DPA) from the spores. Ca^{2+}-DPA is a spore core specific component that is released during spore germination and that triggers activation of specific cortex hydrolases in the cortex to complete germination.

A major obstacle for applying this two-step process for inactivating spores by mild high pressure treatments is that a considerable fraction of the spore population is more resistant to germination and cannot be inactivated without recourse to harsh (heat) treatments. It is thus of interest to see whether this fraction of so-called super-dormant spores can be reduced by appropriate combinations of germinants in conjunction with high pressure. Therefore, we studied the effect of 60 mM Ca^{2+}, 60 mM

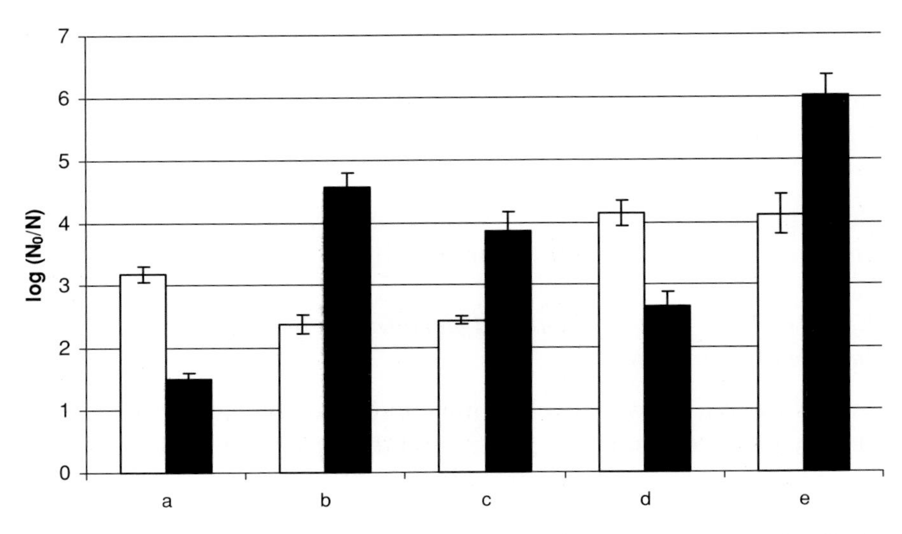

Figure 4.8. Germination of *B. subtilis* PS832 spores by pressure treatment for 60 min at 40°C at 100 MPa (open bar) and 600 MPa (filled bar) in Tris-HCl buffer (100 mM, pH 7.0) by pressure alone (a), by heat pretreatment (80°C/30 min) followed by pressure treatment (b), and by pressure treatment in the presence of, respectively, 60 mM DPA (c), 60 mM Ca^{2+} (d), and 60 mM Ca^{2+} DPA (e). N_0 represents the plate count of the untreated spore suspension and N the plate count after pressure treatment in the case of inactivation, and the plate count after pressure and heat pretreatment (80°C, 30 min).

DPA, and 60 mM Ca^{2+}-DPA on the germination of *B. subtilis* spores in response to pressure (100 or 600 MPa at 40°C for 60 min), and also the effect of a heat shock pretreatment (80°C for 30 min), which is known to activate *B. subtilis* spores and increase their germination rate under pressure (Keynan and Evenchick, 1969).

Figure 4.8 shows that a heat pretreatment at 80°C/30 min, while not affecting spore viability (data not shown), significantly increases germination at 600 MPa. An extra 3 log units of germinated *B. subtilis* spores were detected compared to germination by pressure treatment alone. In contrast, the same heat pretreatment reduced the germination at 100 MPa by 1 log unit. Thus, heat pretreatment stimulates the Ca^{2+}-DPA release induced by high pressures ($>$ 300 MPa).

Further, the presence of Ca^{2+}, DPA, and Ca^{2+}-DPA all increased germination, both at 100 MPa and at 600 MPa. The strongest effect was seen with Ca^{2+}-DPA, showing respective increases of 1.0 and 4.5 log units. It follows from these results that, while 600 MPa is known to induce spore germination via the release of Ca^{2+}-DPA, the addition of more exogeneous Ca^{2+}-DPA can further enhance germination. Since exogeneous Ca^{2+}-DPA germinated 1.0 log unit of *B. subtilis* spores within 60 min after addition, the combination of exogeneous Ca^{2+}-DPA with pressure at 100 MPa was additive, while the combination with 600 MPa increased germination synergistically. Furthermore, the finding that specific germinants can increase the germination of *B. subtilis* spores by pressure opens the possibility of the use of treatments combining high pressures, heat, and/or germinants for food preservation purposes. Some foods may contain intrinsically high levels or a variety of germinants for different types of spores, and thus induce high levels of spore germination by high pressure, while in other foods, the addition of specific germinants may be considered.

High Pressure Induced Germination and Inactivation in Milk

We compared the high pressure germination and inactivation of *B. cereus* spores in milk versus buffer, in order to see whether naturally present germinants in milk would have a stimulatory effect, as anticipated from the previous section (Van Opstal et al., 2004). *Bacillus cereus* is a ubiquitous foodborne pathogen that can be found in a wide range of raw food materials including milk. Thermal pasteurization of milk is insufficient to reduce the numbers of this pathogenic spore, with decimal reduction times at 100°C of 2.2–5.4 minutes. In addition, some *B. cereus* strains can grow at refrigerated temperatures as low as 4–5°C (Choma et al., 2000), making this pathogen an important concern for the safety of pasteurized low-acid foods including milk. Under specific conditions, mild heat treatment might even activate rather than inactivate the dormant *B. cereus* spores, thereby increasing the risk of pathogen outgrowth and food poisoning (Kim and Foegeding, 1990).

Studies on *B. cereus* spore germination/inactivation are limited in the literature. Oh and Moon (2003) reported that germination and inactivation did not exceed 2 logs at 20°C, irrespective of the pressure, but reached 4–6 logs at 600 MPa/40°C, and up to 6–7 logs at 600 MPa/60°C, depending on the sporulation medium used to produce the spores. Similar results were also obtained previously in McIlvaine buffers by Raso et al. (1998), who reported up to 3.2 logs germination and 0.4 logs inactivation after treatment for 15 min at 250 MPa/25°C, and up to 6.9 logs

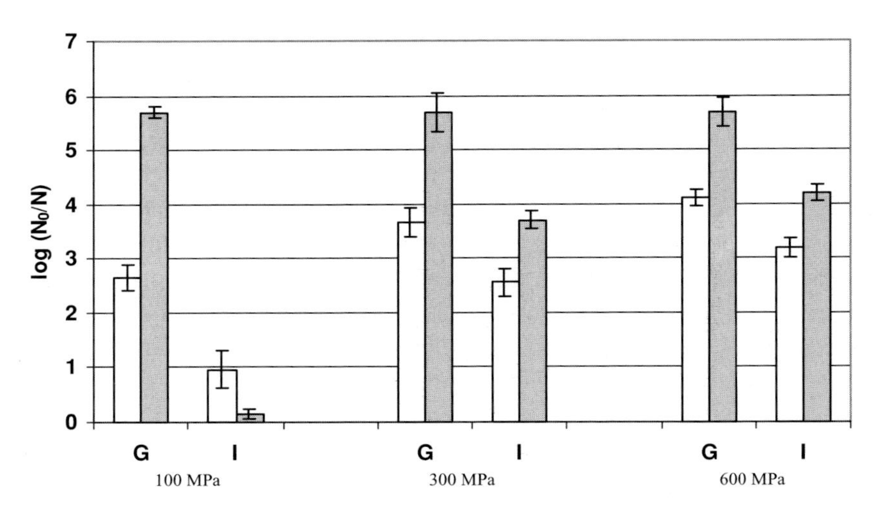

Figure 4.9. Germination (G) and inactivation (I) of *B. cereus* LMG6910 spores by pressure treatment at, respectively, 100, 300, or 600 MPa for 30 min at 40°C in potassium phosphate buffer (100 mM, pH 6.7) (open bar) and milk (grey bar). N_0 represents the plate count of the untreated spore suspension, and N the plate count after pressure treatment in the case of inactivation, and the plate count after pressure and heat treatment (60°C, 10 min) in the case of germination.

germination and 5 logs inactivation after treatment for 15 min at 690 MPa/40°C. McClements et al. (2001) treated two *B. cereus* strains in milk at 400 MPa, and found no inactivation at 8°C and 0.5 log reduction at 30°C. Shearer et al. (2000) reported a 4 log inactivation of *B. cereus* spores after a 10-min treatment at 392 MPa/45°C. Together, these results suggest that pressure treatment has some potential for reducing the viability of *B. cereus* spores in foods, but substantially more work is needed to identify the most efficient process conditions for achieving this inactivation, particularly in real food systems.

We subjected *B. cereus* LMG6910 spore suspensions (inocula of 6×10^6 cfu mL^{-1}) in skim milk (pH 6.7) or in 100 mM potassium phosphate buffer (pH 6.7) to different constant high pressures (100, 300, and 600 MPa) at 40°C for 30 min, and determined the extent of spore germination and inactivation (Figure 4.9). At ambient pressure, 1 log (i.e., 90%) of *B. cereus* spores germinated after 30 min at 40°C in skim milk due to the presence of intrinsic germinants. In phosphate buffer, no germination was detected at this temperature (data not shown). The application of pressure significantly enhanced germination of the spores, as assessed by heat sensitivity (80°C for 10 min). In milk, almost the entire spore population (6 logs) germinated irrespective of the pressure level, while in the phosphate buffer, the extent of germination was incomplete, increasing with pressure from 2.5 logs at 100 MPa to 4 logs at 600 MPa. Spore inactivation also occurred as a result of the pressure treatments, and the levels of inactivation increased with the applied pressure. In all cases, the extent of inactivation was lower than that of germination, indicating that not all of the germinated spores could be inactivated by high pressure, even at 600 MPa. Again, this highlights that the population of germinated spores at 600 MPa is heterogeneous and features different levels of resistance to high pressures, as already pointed out in the section "Stability of Gfp in Spores and Pregerminated Spores at Low pH," where we observed that some, but not all, of the spores that germinated at 600 MPa were inactivated at low pH.

We performed a more detailed study of pressure-induced germination and inactivation of *B. cereus* spores in skim milk to identify candidate treatments capable of ensuring a 5–6 log inactivation of *B. cereus* spores. Spore suspensions were subjected to different combinations of pressure (100–600 MPa) and temperature (30–60°C) for a processing time of 30 min. At ambient pressure, germination increased with temperature from 0.5 logs at 30°C to 2 logs at 60°C in 30 min (data not shown). The level of germination occurring in response to pressure treatments (Figure 4.10A) increased at all temperatures from 100 MPa to 200 MPa, but remained relatively constant at pressures $>$200 MPa. Furthermore, at all pressures, the level of germination increased as the temperature increased from 30 to 40°C, but showed much less or no increase at temperatures $>$ 40°C. Taken together, all treatments at $\geq$200 MPa and $\geq$40°C induced 6 to almost 8 logs of germination, while pressure treatments at 30°C induced only 3–5 logs of germination. The corresponding pressure-induced inactivation of *B. cereus* spores is shown in Figure 4.10B. No inactivation occurred with a treatment at 100 MPa at the lowest temperatures (30, 40, and 45°C). For the other treatments, the inactivation levels increased with temperature at all pressures. The inactivation levels generally also increased with pressure, except for pressures $\geq$300 MPa at 30°C and $\geq$200 MPa at 40°C. A comparison of Figures 4.10A and 4.10B shows that inactivation never reached the same level as germination. Nevertheless, $\geq$6 logs of inactivation were achieved by treatments of 60°C and pressures $\geq$400MPa. Alternatively, based on the germination levels, we

a

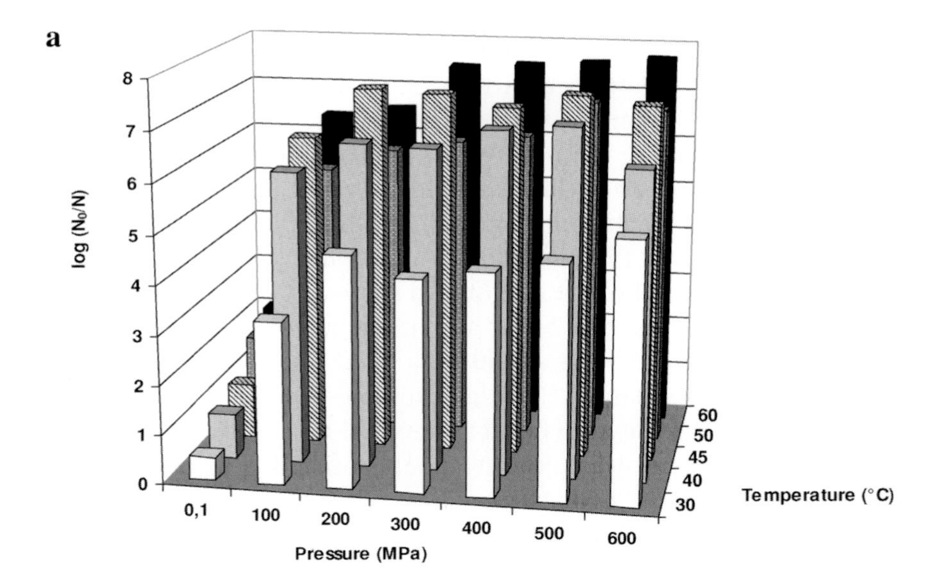

b

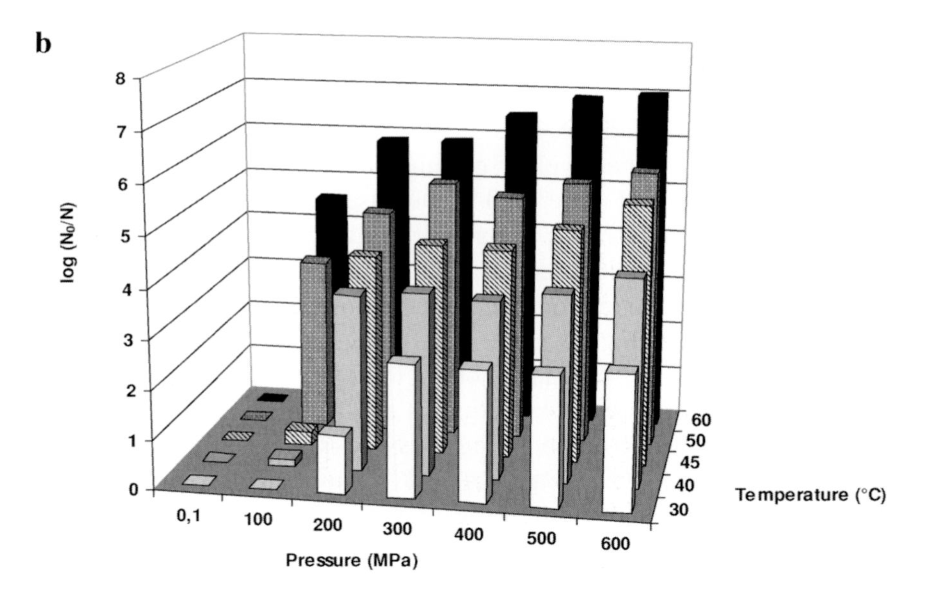

Figure 4.10. Germination (A) and inactivation (B) of *B. cereus* LMG6910 spores in milk by pressure treatment for 30 min at 100–600 MPa and at 30°C, 40°C, 45°C, 50°C, and 60°C. N_0 represents the plate count of the untreated spore suspension (6×10^8 cfu mL^{-1} in this experiment). N is the plate count after pressure treatment and a subsequent heat treatment (60°C, 10 min) to kill the germinated fraction (A), or after a pressure treatment without subsequent heating (B).

Table 4.1. High pressure inactivation of four *B. cereus* strains in milk. N_0 represents the plate count of the untreated spore suspension and N the plate count after pressure treatment

		Logarithmic viability reduction ($\log N_0/N$) ± SD)	
		Pressure treatment time	
B. cereus strain	Treatment[a]	15 min	30 min
LMG6910	A	5.4 ± 0.3	6.7 ± 0.3
	B	5.1 ± 0.2	7.4 ± 0.1
INRAAV TZ415	A	6.6 ± 0.2	6.7 ± 0.2
	B	5.4 ± 0.2	6.3 ± 0.1
INRAAV P21S	A	5.6 ± 0.1	5.8 ± 0.1
	B	6.8 ± 0.2	7.2 ± 0.3
INRAAV Z4222	A	5.6 ± 0.2	6.2 ± 0.1
	B	6.1 ± 0.1	6.4 ± 0.1

[a]A: Pressure treatment at 500 MPa/60°C for, respectively, 15 or 30 min.
B: Pressure treatment of 200 MPa, 45°C for, respectively, 15 or 30 min followed by heat treatment at 60°C for 10 min.
Reprinted from Van Opstal et al., 2004, with permission from Elsevier.

anticipate that a 6 log inactivation can also be achieved by inducing a 6 log germination by high pressure first, and subsequently applying a mild heat treatment (e.g., 60°C for 10 min) to kill the germinated spores. A pressure treatment of 200 MPa at 40°C would suffice for this purpose.

Because the effects of high pressures on microorganisms tend to vary with the strain, we repeated two treatments that should give a 6-D reduction with three different *B. cereus* strains. The first treatment consisted of exposure to 500 MPa/60°C for 30 min, and the second treatment consisted of a sequential process involving exposure to 200 MPa/45°C for 30 min followed by a 10-min treatment at 60°C. An additional experiment was conducted using the identical pressure and temperature conditions, but reducing the treatment time to 15 min, which is a processing time more suited for industrial application (Table 4.1).

The 30-min pressure treatment at 500 MPa/60°C caused $>$ 6 log inactivation of all but one strain, which remained slightly below that value (5.8 logs). With a treatment time of 15 min, only one strain could be inactivated > 6 logs, but the inactivation levels for the other strains were ≥5.4 logs. For the sequential process (high pressure followed by heat treatment), all strains were reduced by ≥6.3 logs. In this case, when the pressure treatment time was reduced to 15 min, the extent of inactivation ranged from 5.1 to 6.8 logs, depending on the strain.

Conclusions

It follows from these results that *B. cereus* spores germinate efficiently in skim milk under high pressure, probably because pressure and intrinsic germinants in milk act in concert to synergistically induce spore germination. Two different processing conditions have demonstrated the capacity to reduce 5–6 logs of four different strains of *B. cereus* spores in milk. One approach is a single-step pressure treatment at 500 MPa/60°C, and the other is a milder, two-step treatment consisting of pressurization at 200 MPa/45°C, followed by heating at 60°C. These results show that mild pressure treatments at ≤600 MPa and ≤ 60°C can contribute significantly to ensuring the safety of minimally processed foods with respect to *B. cereus* spores, and open prospective methods for controlling other sporeformers, pathogenic bacteria, and food spoilage organisms in foods.

Acknowledgments

Authors I.V. and A.A. were supported by a post-doctoral fellowship of, respectively, the K.U. Leuven Research Council and the Fund for Scientific Research Flanders (F.W.O.).

References

Bender, G.R., and R.E. Marquis. 1982. Sensitivity of various salt forms of *Bacillus megaterium* spores to the germination action of hydrostatic pressure. *Canadian Journal of Microbiology* 28:643–649.

Black, E.P., K. Koziol-Dube, D.S. Guan, H. Wei, B. Setlow, D.E. Cortezzo, D.G. Hoover, and P. Setlow. 2005. Factors influencing germination of *Bacillus subtilis* spores via activation of nutrient receptors by high pressure. *Applied and Environmental Microbiology* 71:5879–5887.

Bokman S.H., and W.W. Ward. 1981. Renaturation of aequorea green-fluorescent protein. *Biochem. Biophys. Res. Commun.* 101:1372–41380.

Choma, C., T. Clavel, H. Dominguez, N. Razafindramboa, H. Soumille, C. Nguyen-the, and P. Schmitt. 2000. Effect of temperature on growth characteristics of *Bacillus cereus* TZ415. *International Journal of Food Microbiology* 55:73–77.

Clouston, J.G., and P.A. Wills. 1969. Initiation of germination and inactivation of *Bacillus pumilus* spores by hydrostatic pressure. *Journal of Bacteriology* 97:684–690.

Crameri A., E.A. Whitehorn, E. Tate, and W.P.C. Stemmer. 1996. Improved green fluorescent protein by molecular evolution using DNA shuffling. *Nat. Biotechnol.* 14:315–319.

Cubitt, A.B., R. Heim, S.R. Adams, A.E. Boyd, L.A. Gross, and R.Y. Tsien. 1995. Understanding, improving and using green fluorescent proteins. *Trends in Biochemical Sciences* 20:448–455.

Elsliger, M.-A., R.M. Wachter, G.T. Hanson, K. Kallio, and S.J. Remington. 1999. Structural and spectral response of green fluorescent protein variants to changes in pH. *Biochemistry* 38:5296–5301.

Fernandez P.S., F.J. Gomez, M.J. Ocio, M. Rodrigo, T. Sanchez, and A. Martinez. 1994. D value of *Bacillus stearothermophilus* spores as a function of pH and recovery medium acidulant. *Journal of Food Protection* 58:628–632.

Hanson, G.T., T.B. McAnaney, E.S. Park, M.E. Rendell, D.K. Yarbrough, S. Chu, L. Xi, S.G. Boxer, M.H. Montrose, and S.J. Remington. 2002. Green fluorescent protein variants as ratiometric dual emission pH sensors. 1. Structural characterization and preliminary application. *Biochemistry* 41:15477–15488.

Heim, R., A. Cubitt, and R. Tsien. 1995. Improved green fluorescence. *Nature* 373:663–664.

Igura, N., Y. Kamimura, M.S. Islam, M. Shimoda, and I. Hayakawa. 2003. Effects of minerals on resistance of *Bacillus subtilis* spores at heat and hydrostatic pressure. *Applied and Environmental Microbiology* 69:6307–6310.

Keynan, A., and Z. Evenchick. 1969. “Activation.” In: *The Bacterial Spore*, ed. G.W. Gould and A. Hurst, pp. 259–396. London: Academic Press.

Kim, J., and P.M. Foegeding. 1990. Effects of heat-treatment, $CaCl_2$-treatment and ethanol-treatment on activation of *Bacillus* spores. *Journal of Applied Bacteriology* 69:414–420.

Kneen, M., J. Farinas, Y. Li, and A.S. Verkman. 1998. Green fluorescent protein as a noninvasive intracellular pH indicator. *Biophysical Journal* 74:1591–1599.

Lewis, P.J., and A.L. Marston. 1999. GFP vectors for controlled expression and dual labeling of protein fusions in *Bacillus subtilis*. *Gene* 227:101–109.

Llopis, J., J.M. McCaffery, A. Miyawaki, M.G. Farquhar, and R.Y. Tsien. 1998. Measurement of cytosolic, mitochondrial, and Golgi pH in single living cells with green fluorescent proteins. *Proceedings of the National Academy of Science USA* 95:6803–6808.

Lun, S., and P.J. Willson. 2004. Expression of green fluorescent protein and its application in pathogenesis studies of serotype 2 *Streptococcus suis*. *Journal of Microbiological Methods* 56:401–412.

McClements, J.M.J., M.J. Patterson, and M. Linton. 2001. The effect of growth stage and growth temperature on high hydrostatic pressure inactivation of some psychotropic bacteria in milk. *Journal of Food Protection* 64:514–522.

Miyawaki, A., J. Llopis, R. Heim, J.M. McCaffery, J.A. Adams, M. Ikura, and R.Y. Tsien. 1997. Fluorescent indicators for Ca^{2+}based on green fluorescent proteins and calmodulin. *Nature* 388:882–887.

Moir, A. 2003. Bacterial spore germination and protein mobility. *Trends in Microbiology* 11:452–454.

Oh, S., and M.J. Moon. 2003. Inactivation of *Bacillus cereus* spores by high hydrostatic pressure at different temperatures. *Journal of Food Protection* 66:599–603.

Raso, J., M.M. Gogora-Nieto, G. Barbosa-Canovas, and B.G. Swanson. 1998. Influence of several environmental factors on the initiation of germination and inactivation of *Bacillus cereus* by high hydrostatic pressure. *Int. J. Food Microbiol.* 44:125–132.

Romoser, V.A., P.M. Hinkle, and A. Persechini. 1997. Detection in living cells of Ca^{2+}-dependent changes in the fluorescence emission of an indicator composed of two green fluorescent protein variants linked by a calmodulin-binding sequence. A new class of fluorescent indicators. *Journal of Biological Chemistry* 272:13270–13274.

Sale, A.J.H., G.W. Gould, and W.A. Hamilton. 1970. Inactivation of bacterial spores by hydrostatic pressure. *Journal of General Microbiology* 60:323–334.

Setlow, P. 2003. Spore germination. *Current Opinion in Microbiology* 6:550–556.

Shearer, A.E.H., C.P. Dunne, A. Sikes, and D.G. Hoover. 2000. Bacterial spore inhibition and inactivation in foods by pressure, chemical preservatives, and mild heat. *Journal of Food Protection* 63:1503–1510.

Swerdlow, B.M., B. Setlow, and P. Setlow. 1981. Levels of H^+ and other monovalent cations in dormant and germinating spores of *Bacillus megaterium*. *Journal of Bacteriology* 148:20–29.

Terskikh, A., A. Fradkov, G. Ermakova, A. Zaraisky, P. Tan, A.V. Kajava, X. Zhao, S. Lukyanov, M. Matz, S. Kim, I. Weissman, and P. Siebert. 2000. "Fluorescent timer": Protein that changes color with time. *Science* 290:1585–1588.

Van Opstal, I., C.F. Bagamboula, S.C.M. Vanmuysen, E.Y. Wuytack, and C.W. Michiels. 2004. Inactivation of *Bacillus cereus* spores in milk by mild pressure and heat treatments. *International Journal of Food Microbiology* 92:227–234.

Webb, C.D., A. Decatur, A. Teleman, and R. Losick. 1995. Use of green fluorescent protein for visualization of cell-specific gene expression and subcellular protein localization during sporulation in *Bacillus subtilis*. *Journal of Bacteriology* 177(20):5906–5911.

Wuytack, E.Y., S. Boven, and C.W. Michiels. 1998. Comparative study of pressure-induced germination of *Bacillus subtilis* spores at low and high pressures. *Applied and Environmental Microbiology* 64:3220–3224.

Wuytack, E.Y., and C.W. Michiels. 2001. A study on the effects of high pressure and heat on *Bacillus subtilis* spores at low pH. *International Journal of Food Microbiology* 64:333–341.

Wuytack, E.Y., J. Soons, F. Poschet, and C.W. Michiels. 2000. Comparative study of pressure- and nutrient-induced germination of *Bacillus subtilis* spores. *Applied and Environmental Microbiology* 66:257–261.

Chapter 5

Pressure and Heat Resistance of *Clostridium botulinum* and Other Endospores

Michael G. Gänzle, Dirk Margosch, Roman Buckow, Matthias A. Ehrmann, Volker Heinz, and Rudi F. Vogel

Introduction

The preservation of low-acid canned food to achieve non-refrigerated, shelf-stable products requires the elimination of vegetative microbial cells as well as the inactivation of bacterial endospores capable of growing in the product at the conditions prevailing during distribution and storage. To achieve "commercial sterility," dormant bacterial endospores require attention because they are much more resistant than vegetative bacterial cells, viruses, yeasts, and fungi to physical and chemical methods of food preservation. Endospores are resistant to a wide range of bactericidal treatments such as wet and dry heat, high pressure, solvents, disinfectants, UV light, and ionizing radiation. This high level of spore resistance is attributed to the low water content of the spore core, which is thought to be in a glassy state in dormant spores (Ablett et al., 1999); the cortex structure, which is characterized by a thick peptidoglycan layer and a low permeability to hydrophilic molecules; the high level of minerals (e.g., Ca^{2+} ions) and dipicolinic acid (DPA) in the core; and the activity of DNA-repair mechanisms during germination (Nicholson et al., 2000; Atrih and Foster, 2002; Ablett et al., 1999).

Currently, the commercial sterility of low-acid canned foods is achieved by conventional thermal processing. In these cases, the organism that is the primary target of destruction to achieve food safety is *Clostridium botulinum,* a species of phylogenetically and physiologically diverse anaerobic sporeforming rods that produce potent botulinal neurotoxins during growth (Collins and East, 1998). The elimination of 12 log units (12D) *Clostridium botulinum* is considered to require a minimum treatment for 2.4 min at 121°C. To inactivate spores of heat resistant mesophilic spoilage organisms (e.g., *C. sporogenes* with a $D_{121^\circ C}$-value of 1 min), the treatment intensity is usually equivalent to a treatment of at least 5 min at 121°C. The preservation of canned food intended for distribution in tropical climates requires higher treatment intensities to eliminate *Geobacillus stearothermophilus* (formerly *Bacillus stearothermophilus*) and *Thermoanaerobacterium thermosaccharolyticum* (formerly *Clostridium thermosaccharolyticum*), both of which are thermophilic organisms with $D_{121^\circ C}$-values ranging from 3 to 10 min.

The long treatment times at high temperatures that are required to achieve commercial sterility cause physical and chemical changes that compromise food quality, especially for solid or highly viscous foods. Heat transfer takes place predominantly through conduction. High pressure processing (HPP) is currently investigated as a possible alternative to thermal processes at ambient pressure to inactivate bacterial endospores in low-acid canned foods. Two processes employing high hydrostatic pressure are currently investigated, the pressure-induced germination of spores, and high pressure/high temperature (HP/HT) processes to achieve the direct inactivation of bacterial endospores. Pressure in the range of 100–600 MPa at ambient temperature induces spore germination, and the germinated spores are eliminated by the subsequent application of pressure at ambient temperature, or thermal pasteurization (Herdegen, 1998; Wuytack et al., 1998; Paidhungat et al., 2002). Studies on the population heterogeneity of bacilli and *C. botulinum* spores have shown that a small proportion of the spores remain dormant (Margosch et al., 2004a; Stringer et al., 2005), and therefore even repeated pressure cycles fail to achieve commercial sterility of foods. Some spores can tolerate up to 1,500 MPa at ambient temperature without losing their viability (Rovere et al., 1998). The application of high pressure at pasteurization temperatures inactivates most spores that tolerate pasteurization temperatures at ambient pressure without losing their viability (Sale et al., 1970; Rovere et al., 1998; San Martín et al., 2002).

Changes in pressure are thermodynamically coupled to changes in temperature. Compression heating and decompression cooling may be exploited to achieve rapid and uniform adiabatic heating and cooling of the product. Based on the assumption that the heat resistance of bacterial endospores at high pressure conditions is equal to or less than their resistance at ambient conditions, the compression heating may be used to significantly reduce the heating and cooling times compared to conventional thermal sterilization processes (de Heij et al., 2003).

Although the effects of pressure on bacterial endospores have been studied for almost a century, information on the pressure resistance of *C. botulinum* as a prerequisite for the commercialization of pressure processes in food preservation has become available only in recent years (Rovere et al., 1998; Reddy et al., 1999, 2003, 2006; Margosch et al., 2004b, 2006; Gola and Rovere, 2005). In this chapter, an overview is given of the current knowledge on the pressure-induced inactivation of *C. botulinum* spores as a safety determinant in low-acid canned foods, and spores from other organisms relevant to food spoilage.

Selection of Pressure-Resistant Bacterial Endospores I: Influence of Sporulation Conditions

Knowledge of the effect of sporulation conditions on the pressure resistance of endospores helps establish the most resistant types of spores from a relevant organism that can be used as an indicator of a "worst-case scenario." Spore mineralization, as well as the appropriate ratio of minerals, influences spore resistance to wet heat (Atrih and Foster, 2002; Nicholson et al., 2000). Spores contain high levels of divalent cations (Ca^{2+}, Mg^{2+}, and Mn^{2+}), and the mineral content of the spore can be manipulated by adjusting the mineral content of the sporulation medium. *B. subtilis* spores formed at high sporulation temperatures and high mineral contents in the sporulation medium had decreased resistance to pressure but increased resistance to heat (Margosch et al., 2004a; Igura et al., 2003). The influence of the sporulation medium on pressure resistance was demonstrated for *C. botulinum* spores (Margosch et al., 2004b). Since three of seven *C. botulinum* strains failed to produce spores on standard 1 medium (ST1) or Reinforced Clostridial Medium (RCM), a novel medium including soil extract, termed WSH medium, was formulated. For those strains that formed spores on all three media, the pressure and heat resistance of the spores was compared for the various media. Spores from *C. botulinum* grown on WSH medium exhibited a higher resistance to heat and pressure treatments. For example, treatments for 5 min at 100°C or for 16 min at 600 MPa and 80°C reduced spore counts of WSH-grown *C. botulinum* by 1.2 and 2.9 log units, respectively, whereas the same treatments respectively reduced spore counts of ST1 and RCM spores by more than 3 and 6 log units. The increased resistance of WSH-grown spores may be attributable to the increased supply of minerals that can be derived from the soil extract. Since soil is the likely source of *C. botulinum* spores contaminating foods, WSH-grown spores, accordingly, may be more representative of spores occurring in food.

Selection of Pressure-Resistant Bacterial Endospores II: Identification of Relevant Pressure-Resistant Strains

The validation of high pressure processes for food preservation and ensuring food safety requires studies using the most pressure-resistant strains of *C. botulinum* and the most pressure-resistant strains of relevant food spoilage organisms (Sizer et al., 2002). To facilitate the comparison between heat and pressure resistance of bacterial endospores, pressure-death time curves for spores from selected aerobic and anaerobic bacteria relevant in foods are shown in Figure 5.1. Spores from *B. amyloliquefaciens* TMW 2.479, an isolate from ropy bread, and *C. botulinum* type B TMW 2.357 have relatively low resistances to wet heat in comparison to *Geobacillus stearothermophilus* and *Thermoanaerobacterium thermosaccharolyticum*, but their resistances to pressure applications at 70–80°C are considerably higher (Figure 5.1; Margosch et al., 2004a, 2004b; Ananta et al., 2001). Spores of *C. botulinum* type A and type B are more resistant to pressure than spores of *C. sporogenes* and spores of *G. stearothermophilus* (Rovere et al., 1998; Reddy et al., 2003; Gola and Rovere, 2005). Spores of non-proteolytic, group II *C. botulinum* strains are less pressure resistant than proteolytic *C. botulinum*, and the former are inactivated by more than 4 log units at 800 MPa at 50°C (Reddy et al., 1999, 2006).

A large variation in the pressure resistance of strains from the same species is also apparent in Figures 5.1 and 5.2 for the species *B. subtilis* and *C. botulinum* (Margosch et al., 2004a, 2004b). In these two species,

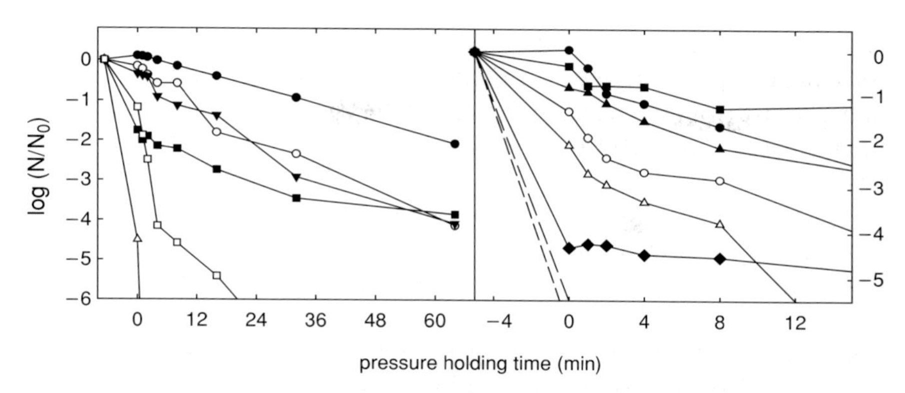

Figure 5.1. Inactivation of endospores of *Bacillus* spp., *Clostridium botulinum*, and *Thermoanaerobacterium thermosaccharolyticum* in mashed carrots by combined application of high pressure and heat. Panel A: Inactivation of *Bacillus* spp. at 800 MPa, 70°C: •, *B. amyloliquefaciens* FAD82 (TMW 2.479); ○, *B. amyloliquefaciens* FAD We (TMW 2.475); ▾, *B. subtilis* FAD109 (TMW 2.483); ■, *B. smithii* TMW 2.487; □, *B. licheniformis* TMW 2.492; △, *B. subtilis* DSM618. Panel B: Inactivation of *C. botulinum* and *T. thermosaccharolyticum* at 600 MPa, 80°C: ■, *C. botulinum* type B, REB89 (TMW 2.357); •, *C. botulinum* type A, REB1750 (TMW 2.356); ▲, *C. botulinum* TMW 2.359; ○, *C. botulinum* type F, REB1072 (TMW 2.358); △, *C. botulinum* type B, ATCC 19397; ♦, *T. thermosaccharolyticum* TMW 2.299. Lines dropping below the x-axis indicate cell counts below the detection level and all cell counts for *C. botulinum* Type B, Nr. 160 (TMW 2.518) and *C. botulinum* ATCC 25765 were below detection limit after treatment. Figure prepared with data from Margosch et al. (2004a, 2004b).

the spore counts obtained from the most pressure-sensitive strains differ by more than 4 log units from the spore counts of the most pressure-resistant strains after treatments at identical conditions. Strains of *C. botulinum* also exhibit a considerable variation with respect to their heat resistance. The *C. botulinum* type B TMW 2.359 strain was the most heat resistant (with a $D_{120^\circ C}$-value of 1.2 min), and the *C. botulinum* type B TMW 2.357 strain was the most pressure resistant (Margosch et al., 2004b).

The resistance of bacterial endospores to HP/HT treatments does not correlate to their resistance to wet heat. Therefore, the large body of thermal death-time data that is available for spores relevant in food processing can not be extrapolated for use in the validation of high pressure preservation of foods. Moreover, strains from the same species may exhibit large variations with respect to their resistance to HPP

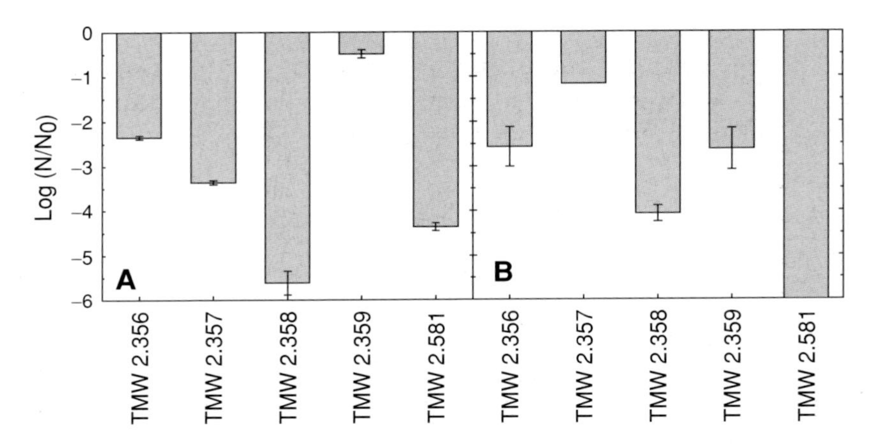

Figure 5.2. Comparison of the heat and pressure resistance of *C. botulinum* strains. Panel A: Cell counts after treatment at 100°C for 10 min in mashed carrots. Panel B: Cell counts after treatment with 600MPa, 80°C for 16 min. Figure prepared with data from Margosch et al. (2004b).

treatments, which necessitates the use of strain cocktails (including relevant food isolates) for the development of HP/HT processes.

Pressure-Induced Loss of DPA and Sublethal Injury of Bacterial Endospores

Bacterial endospores contain about 20% DPA, a dicarboxylic acid that occurs mainly as the corresponding Ca^{2+} salt (CaDPA) in the spore core. CaDPA is a characteristic component of dormant spores that contributes to imparting the spore with extreme resistances to heat and UV-radiation (Paidhungat et al., 2000). The loss of DPA from spores and the concomitant loss of heat resistance is an early event during nutrient-activated spore germination. Treatments at 100 MPa activate spore nutrient receptors independent of the availability of the corresponding nutrient signal, and DPA is released in the physiological pathway called germination (Sale et al., 1970). After treatments of 500–600 MPa, spores become depleted of DPA as a result of altered spore permeability, and germination is induced independent from the nutrient receptors. Pressure-germinated spores can be inactivated by mild heat treatments at 80°C, or by pressure treatments at ambient temperature (Paidhungat et al., 2001, 2002; Wuytack et al., 1998, 2000). The loss of DPA from spores also occurs during lethal treatments of heat (Malidis and Scholefield, 1985) or high pressure (Heinz and Knorr, 1998).

Margosch et al. (2004a, 2004b) compared the loss of DPA from spores of *Bacillus* spp. and *C. botulinum* after exposing these organisms to combined HP/HT treatments. Spores of the pressure-sensitive strains *B. subtilis* TMW 2.485 and *B. licheniformis* TMW 2.492 rapidly lost almost all DPA after treatments with 800 MPa at 70°C, and these DPA-free spores were inactivated readily by subsequent treatments of 80°C. Spores of the pressure-resistant strains of *B. amyloliquefaciens* TMW 2.479 and *C. botulinum* TMW 2.357 had an increased ability to retain DPA during pressure treatments, indicating that the pressure resistance of spores relates to their ability to retain DPA at HP/HT conditions (Margosch et al., 2004a, 2004b). Similarly, the ability of spores to retain DPA during heat treatments has also been suggested to correspond to their heat resistance (Kort et al., 2005). The properties of spore structure or composition that allow spores to retain DPA during high pressure and/or wet heat treatments govern spore resistance, and the retention of DPA by spores may provide a useful probe for rapidly screening the pressure resistance of endospores.

Individual spores of untreated *B. licheniformis* TMW 2.492 grew to detectable turbidity after 12–48 hr, with a median of about 20 hr (Margosch et al., 2004a; Figure 5.3a). Exposure of these spores to mild heat (70°C for 10 min) resulted in rapid germination and outgrowth of individual spores. HPP treatments of the spores with 800 MPa and 70°C for 4 min resulted in a slower process of spore outgrowth, and these HPP-treated spores grew to detectable turbidity only after 20–105 hr with a median of about 60 hr. Spores of the DPA-deficient strain *B. subtilis* CIP 76.26 exhibited longer times-to-detection compared to spores of the same strain that contained DPA (Figure 5.3b; Margosch et al., 2004a), suggesting that the increased times-to-detection after HPP treatments (Figure 5.3a) might in part be attributable to the concomitant loss of DPA. The loss of DPA from spores that is induced by high pressure at high temperatures can be interpreted as a result of the destruction of spore structure(s) and represents sublethal injury rather than a pathway involving spore germination (Margosch et al., 2004a; Heinz and Knorr, 1998).

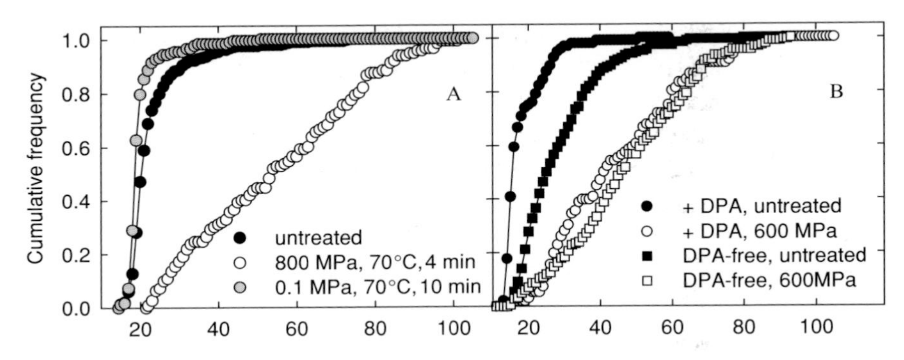

Figure 5.3. Individual detection times of spores of *B. licheniformis* (Panel A) and the DPA-deficient mutant *B. subtilis* CIP 76.26 (Panel B). Spore suspensions of the strains were treated as indicated in the graphs, diluted in ST1 broth to obtain one spore per culture tube and the detection times were determined by periodically measuring the optical density of the culture during incubation at 37°C. The cumulative frequency is plotted based on at least ninety-six observations of individual spores from a sample treated as indicated. The designations +DPA and DPA-free in Panel B refer to DPA-containing and DPA-free spores of the strain as determined by the culture conditions. Figure prepared with data from Margosch et al. (2004a).

HPP-treated spores tend to span a broad distribution of lag times, and viable spore counts may be underestimated when survivors are quantified by standard plate counting procedures (Margosch et al., 2004a). Long incubation times of more than 5 days or the use of most-probable-number techniques are required to enumerate pressure-treated spores, especially in conditions that tend to exhibit long lag phases due to the occurrence of HPP-induced sublethal injuries.

Effect of Pressure and Temperature on the Inactivation of *C. botulinum* Spores

Most studies on the HPP-induced inactivation of spores have been performed at temperatures ranging from 50 to 80°C (i.e., temperatures that do not inactivate endospores at ambient pressure). The high resistance of *C. botulinum* and *B. amyloliquefaciens* spores to HPP necessitated the use of high pressure in conjunction with high temperatures in the range of 90–120°C, a temperature range in which spore inactivation is observed near ambient pressure (Margosch et al., 2006; Gola and

Rovere, 2005). Both the Margosch et al. (2006) and the Gola and Rovere (2005) studies were carried out with high pressure equipment that approximated adiabatic processes and that can be characterized as heating the pre-warmed sample to the desired end-temperature by compression heating, followed by isothermal pressure holding times, then cooling during decompression. Margosch et al. (2006) also compared spore inactivation at HP/HT combinations in the range of 600–1,400 MPa and 70–120°C with thermal inactivation at ambient pressure.

Data for the inactivation of *C. botulinum* type B TMW 2.357 in Tris-His Buffer, pH 5.15 from Margosch et al. (2006) and *C. botulinum* type B (ATCC 25765) in phosphate buffer from Gola and Rovere (2005) are shown in Figure 5.4. Although different strains and buffer systems were used, comparable pressure-death time curves were obtained in these two studies. At temperatures below 90°C and any pressure, *C. botulinum* was inactivated by less than 2 log. In the range of 600–1,400 MPa and 100–120°C, an increase in either pressure or temperature accelerated spore inactivation. However, spore inactivation at 600–800 MPa and 100 or 110°C and was slower than inactivation at the same temperature at ambient pressure. For example, treatments with 600 and 800 MPa at 110°C for 2 min reduced the viable spore counts by about 3 and 4 log, respectively, whereas a more than 5 log reduction was observed after treatments for 2 min at 110°C and 0.1 MPa (Figure 5.4).

High pressure applications in the range of 600–1,000 MPa did not accelerate the thermal inactivation of *G. stearothermophilus* at 130°C (Ardia, 2004), and higher $D_{121°C}$-values were reported for the inactivation of *B. amyloliquefaciens* FAD82 at 500 MPa compared to 0.1 MPa (Rajan et al., 2006). Moreover, *B. amyloliquefaciens* spores exhibited little sensitivity to pressure changes in the range of 600–1,200 MPa at a temperature of 121°C (Margosch et al., 2006; Rajan et al., 2006). Similarly, *T. thermosaccharolyticum* TMW 2.299 tolerated 80°C at ambient pressure but was rapidly inactivated at 200 MPa and 80°C (Figure 5.5). Increasing the pressure level to 600 or 800 MPa did not further increase the inactivation.

It is generally assumed that the inactivation of microorganisms is accelerated by increasing the pressure or temperature. While such a relationship holds for vegetative cells, the present results indicate that spores do not strictly adhere to this relationship at all combinations of high pressure and temperature. Instead, the current data show that

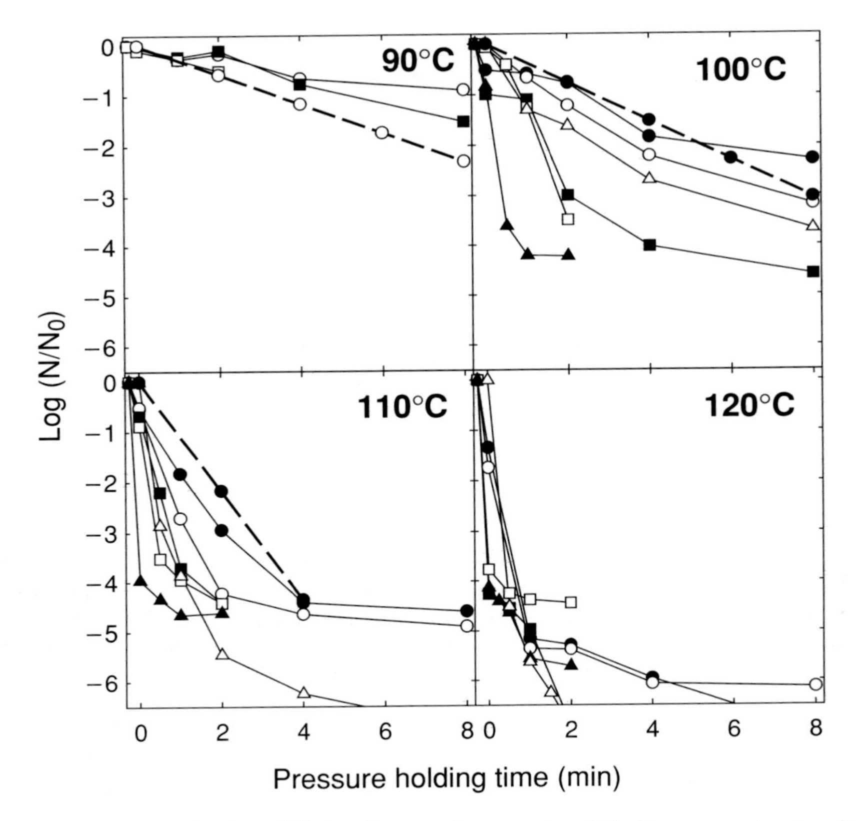

Figure 5.4. Inactivation of *C. botulinum* endospores in adiabatic pressure treatments at 90, 100, 110, and 120°C as indicated. The pressure levels are 0.1 MPa, △; 600 MPa, ●; 800 MPa, ○; 1,000 MPa, ■; 1,200 MPa, □; and 1,400 MPa, ▲. Solid lines indicate the inactivation of *C. botulinum* type B TMW 2.357 in Tris-His buffer, pH 5.15, detection limit: log $(N/N_0) = -6.5$. Figure based on data from Margosch (2004) and Margosch et al. (2006). Dashed lines indicate the inactivation of *C. botulinum* type B ATCC 25765 in phosphate buffer, pH 7.0, detection limit: log $(N/N_0) = -4.5$. Drawing is based on D-values from Gola and Rovere (2005).

pressure contravenes the effects of temperature in some lethal and sublethal regimes.

Pressure-Temperature (P-T) Isokinetics Diagram for the Inactivation of *C. botulinum*

The data set presented by Margosch et al. (2006) covers the combined effects of pressure and temperature on the inactivation of *C. botulinum*

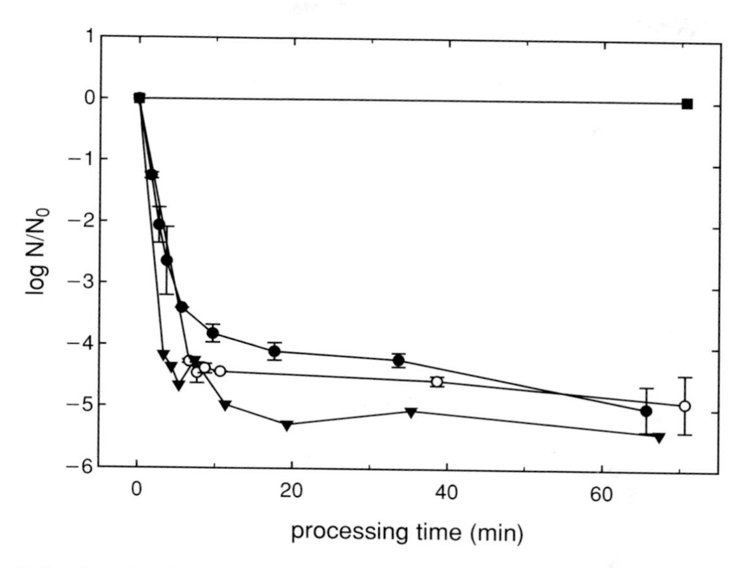

Figure 5.5. Inactivation of *T. thermosaccharolyticum* TMW 2.299 endospores during treatments at 80°C in mashed carrots, pH 5.15. The pressure levels are 0.1 MPa, ■; 200 MPa, •; 600 MPa, ▼; and 800 MPa, ○. Strain cultivation, heat, and pressure treatments were carried out as described by Margosch et al. (2004b).

spores in the parameter ranges of 0.1–1,400 MPa and 60–120°C, and provides the basis for generating a P-T diagram for spore inactivation with isokinetics lines for a 5D reduction of spore counts. A first-order kinetics model is clearly inappropriate for evaluating the pressure-death time curves because of their pronounced nonlinearity. The survivor curves were fitted to an nth order equation:

$$\frac{N}{N_0} = (1 + k't(n-1))^{\frac{1}{1-n}} \qquad \text{with } k' = k(N_0)^{(n-1)} \tag{5.1}$$

with t = time (min) N and N_0 = spore counts at time t and initial spore counts, respectively (cfu/mL), k = rate constant (min^{-1}) and n = reaction order. Analysis of the survivor curves showed that all the experimental data could be fitted when the reaction order had a single fixed value of n = 1.35. The rate constants obtained for each set of experimental conditions were related to P and T using the following quadratic model:

$$\ln(k') = a_0 + a_1 p + a_2 T + a_3 p^2 + a_4 T^2 + a_5 pT + a_6 p^2 T \tag{5.2}$$

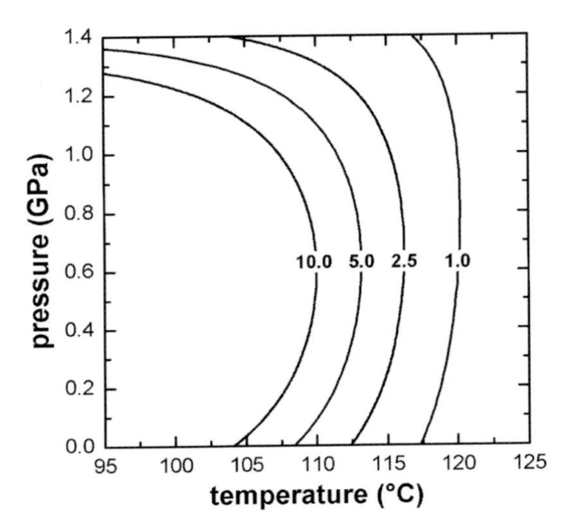

Figure 5.6. Pressure-temperature isokinetics diagram for 5 log reduction of *C. botulinum* TMW 2.357 after 10, 5, 2.5, and 1 min. The figure is drawn based on a model for pressure-temperature inactivation of *C. botulinum* TMW 2.357 in Tris-His buffer, pH 5.15 established by Margosch et al. (2006).

Values of the parameters a_0–a_6 were estimated as:

$$a_0 = 2.465, a_1 = -0.023, a_2 = -0.149, a_3 = 2.259 \times 10^{-5},$$
$$a_4 = 1.462 \times 10^{-3}, a_5 = 1.798 \times 10^{-4}, a_6 = -1.806 \times 10^{-7}$$

This simple model generally predicted the experimental data within a deviation of 1.5 log units (data not shown) and facilitates the presentation of the survival data at the various combinations of pressure and temperature as a single isokinetics diagram (Figure 5.6). The protective effect of high pressure versus thermal inactivation in the range of 600–1,000 MPa, in agreement with experimental data, is apparent in Figure 5.6. Discrepancies between the predicted and experimental values can in part be attributed to the use of a single fixed reaction order to describe all of the survivor curves. Survivor curves with an apparent log-linear shape (e.g., the data at 600 MPa and 100°C in Figure 5.4) can be described more appropriately with a value of $n = 1$, whereas curves exhibiting pronounced tailing (e.g., 800 MPa, 110°C) are better described with values of $n > 1.35$. These deviations can be addressed by using reaction orders (n) that are functions of the variables P and T, or by using a model that takes into account the subpopulations featuring different levels of resistance (Gänzle et al., 2001; Kort et al., 2005).

Either approach would change the overall structure of the model and increase the number of parameters, thereby preventing the representation of all relevant information derived from the various survivor curves in a single diagram.

The elliptic shape of the isokinetics diagram resembles the P,T phase diagrams used to characterize the denaturation of proteins. Indeed, the secondary model describing the inactivation rate constant as a function of P and T was formulated in a manner analogous to mathematical models used to describe the phase boundaries of proteins (Smeller, 2002; Smelt et al., 2002). Phase diagrams have been established for many proteins and the phase boundary lines frequently have two extreme values, a temperature >0°C for maximum baro-resistance of the protein, and a pressure >0.1 MPa for maximum thermoresistance of the protein (for review, see Smeller, 2002). A comparable analogy to phase diagrams of proteins exists for the P,T isokinetics diagrams derived from survivor curves of vegetative bacteria (Smelt et al., 2002). However, it can be expected that there are multiple cellular targets that are responsible for the inactivation of bacterial endospores by the combined application of pressure and heat that can not be fully represented as protein denaturation. The mechanisms of spore inactivation at the extremes of high pressure and temperature are not fully understood, and further studies are required to identify those structural components in bacterial endospores that represent the targets most sensitive to pressure and that render the spores susceptible to inactivation.

Tailing of Pressure-Death Time Curves

The survivor curves at 100–120°C and high pressure conditions of *C. botulinum* type B TMW 2.357 exhibited pronounced tailing, indicating the existence of a subpopulation with enhanced resistance to high pressure. The physiological heterogeneity of individual spores in a population cultured from a clonal spore preparation can have substantial variability in their inherent resistance to lethal agents, and, accordingly, the survival curves can exhibit nonlinearities in the form of tailing. The fraction of more resistant spores in the population was unaffected by the level of high pressure, but did depend on the temperature. Approximately 1 in 10^4, 10^5, and 10^6 of the spore populations were resistant to pressure treatments at 100, 110, and 120°C, respectively.

Tailing has been reported for HP/HT combinations for the inactivation of spores of *Alicyclobacillus acidoterrestris* (Lee et al., 2002), *Geobacillus stearothermophilus* (Ananta et al., 2001), *C. botulinum* (Reddy et al., 2003), *B. amyloliquefaciens* (Rajan et al., 2006), and *B. anthracis* (Cléry-Barraud et al., 2004). Tailing, comparable to that shown in the data in Figure 5.4, has also been observed for thermal death-time curves and was also attributed to the phenotypic heterogeneity of spore populations, rather than genotypic variation (Han, 1975; Cerf, 1977; Kort et al., 2005). The heterogeneity of spores from a clonal population with respect to their experimentally observed detection times is shown in Figure 5.3 (Margosch et al., 2004a). More pronounced tailing was observed in the survivor curves of *C. botulinum* from HP/HT treatments than from heat inactivation at ambient pressure (Figure 5.4).

Effect of pH and Process Conditions on the Inactivation of *C. botulinum*

The composition of the substrate in which the microorganisms are dispersed has a significant influence on the bactericidal effect of the high pressure treatment. The acidity, the concentrations of sugars and salts, and the presence of antibacterial agents can exert substantial synergistic or antagonistic effects on the pressure-mediated elimination of vegetative bacteria (Alpas et al., 2000; Molina-Gutierrez et al., 2002; Kalchayanand et al., 1998). Similarly, low pH, the presence of nisin, or the presence of sucrose laureate increase the effectiveness of high pressure treatments on bacterial spore inactivation (Roberts and Hoover, 1996; Stewart et al., 2000). Conversely, cocoa mass with a water content of 10% or less protected *G. stearothermophilus* spores from the lethal effects of high pressure and temperature (Ananta et al., 2001).

Wuytack and Michiels (2001) suggested that pressure treatments of spores at low pH promote the exchange of minerals from the spore core with protons and decrease the heat resistance of the spores. Margosch et al. (2004b) determined the effect of pH on the pressure-induced inactivation of *C. botulinum* in a pressure-stable Tris-buffer. In this case, the inactivation kinetics were not affected by a pH shift from 6.0 to 5.15. Decreasing the pH to 4.0 accelerated the reduction of viable spore counts (Figure 5.7) and the release of DPA from the spores (Margosch et al., 2004b). The dissociation constants of weak acids (pK_a) are pressure dependent, and high pressure favors the dissociation of phosphate

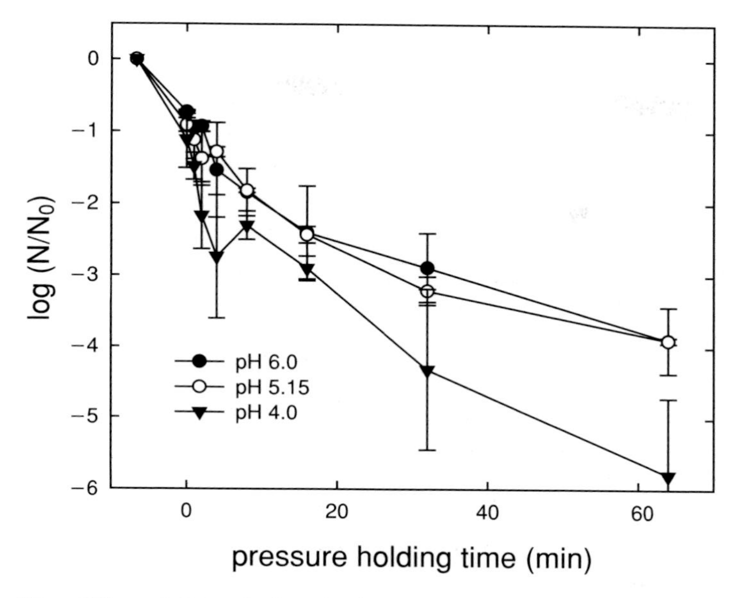

Figure 5.7. Effect of pH on the inactivation of *C. botulinum* TMW 2.357 in Tris-His buffer at 800 MPa, 80°C. Figure adapted from Margosch et al. (2004b).

and carboxylic acids based on their large negative volumes of dissociation (Owen and Brinkley, 1941). Foods in which such compounds act as the primary buffering agent typically experience a reduction in pH by about 1.0 pH units upon compression to 300 MPa (Molina-Gutierrez et al., 2002). Foods with a pH > 4.5 at ambient pressure also experience low pH environments in high pressure conditions.

The inactivation of *C. botulinum* was slightly accelerated in mashed carrots compared to Tris-His buffer. This difference was attributed to changes in the pH experienced in mashed carrots during pressure treatments, since the pH in Tris-His buffer is essentially pressure stable (Margosch et al., 2004b). Reddy et al. (2003) observed a lower rate of inactivation of *C. botulinum* type A spores inoculated in crabmeat compared to the rate of inactivation of these spores when suspended in phosphate buffer.

"Pressure-Assisted Thermal Processing" of Foods: Summary and Perspectives

Historically, the species description of *C. botulinum* included all anaerobic sporeforming rods capable of producing the botulinum neurotoxin.

The application of current concepts in bacterial taxonomy has shown that the species *C. botulinum* forms four distinct phylogenetic lineages related to *C. sporogenes*, *C. butyricum*, *C. novyi*, and *C. subterminale*. Moreover, botulinum neurotoxin-producing strains of *C. baratii* and *C. butyricum* are recognized (Collins and East, 1998).

The 12D concept is based on a $D_{121^\circ C}$-value of 0.204 min for *C. botulinum*. Based on the phylogenetical and physiological heterogeneity of the species *C. botulinum*, individual strains that exhibit a significantly higher heat resistance exist. For example, Margosch et al. (2004b) described a strain of *C. botulinum* type B with a $D_{120^\circ C}$-value of 1.2 min, which is essentially identical to the corresponding D-value of *C. sporogenes*. Based on the $D_{120^\circ C}$-value of the most heat-resistant strain of *C. botulinum*, thermal treatments equivalent to 5 min at 121°C that are currently applied in industrial practice will result in a less than 5D reduction of such spores. The 12D concept or "botulinal cook" has nevertheless virtually eliminated the occurrence of *C. botulinum* in commercially processed canned foods.

Sizer et al. (2002) outlined several approaches for the validation of high pressure processes for low-acid canned foods to ensure a comparable level of safety is achieved. The data that are currently available for the pressure-induced inactivation of *C. botulinum* endospores allow a preliminary assessment of the suitability of high pressure processes for the shelf-stable preservation of low-acid canned foods. Pressure levels of up to 1,500 MPa were applied in laboratory scale trials, but large-scale high pressure equipment is currently limited to pressures in the vicinity of 800 MPa. In the pressure range of 0.1–1,400 MPa, the inactivation of *C. botulinum* spores by more than 5 log requires temperatures of 100°C or higher. Thus, high pressure processes are likely to operate in a temperature range comparable to conventional thermal processing (retorting). The benefit of high pressure processing is that it does not rely merely on a reduction of the process temperature, but on a reduced thermal load that is achieved by substantially accelerated heating and cooling rates.

The heat resistance of spores at ambient pressure does not correlate to their heat resistance at high pressure conditions. Accordingly, kinetics data derived from thermal treatments at ambient pressure can not be used to validate high pressure processes. Additionally, the assumption that the heat resistance of spores at high pressure conditions is reduced compared to their heat resistance at ambient pressure is contravened by observations described above of a protective effect of pressure versus applied heat at certain combinations of pressure and temperature. The occurrence of such a protective effect under some conditions might raise questions regarding the suitability of high pressure applications for the inactivation of *C. botulinum* spores in low-acid canned foods, especially given the success of the traditional 12D concept on which current thermal processes are based. However, pressure-assisted thermal treatments reduce *C. botulinum* contamination in a magnitude equal to that of the thermal processes that are currently applied in industrial practice. Pressure-assisted thermal treatments are therefore suitable for the production of microbiologically safe foods with superior texture and flavor.

Spores of *C. botulinum* are suggested to be among the most pressure-resistant species of spores studied so far. A non-pathogenic or toxinogenic surrogate organism with a significantly higher resistance to pressure at all HP/HT combinations has not yet been identified. Based on studies in the high pressure range of 400–800 MPa and the temperature range of 60–116°C (Margosch et al., 2004b), the strain *B. amyloliquefaciens* TMW 2.479 was suggested as a suitable surrogate organism. This strain exhibited a higher resistance to pressure compared to the most pressure-resistant *C. botulinum* spores and virtually all other spores for which literature data is available. Subsequent studies in the high pressure range of 600–1,400 MPa at temperatures ranging from 60 to 120°C have substantiated this finding (Margosch et al., 2006; Rajan et al., 2006), but caution must be noted. At some combinations of high pressure and temperature, the resistance of *C. botulinum* equals or exceeds that of *B. amyloliquefaciens* (Margosch et al., 2006). Accordingly, the validation of high pressure processes for food preservation therefore requires the use of pressure-resistant *C. botulinum* spores.

References

Ablett, S., A.H. Darke, P.J. Lillford, and D.R. Martin. 1999. Glass formation and dormancy in bacterial spores. *International Journal of Food Science and Technology* 34:59–69.

Alpas, H., N. Kalchayanand, F. Bozoglu, and B. Ray. 2000. Interactions of high hydrostatic pressure, pressurization, temperature and pH on death and injury of pressure-resistant and pressure-sensitive strains of foodborne pathogens. *International Journal of Food Microbiology* 60:33–42.

Ananta, E., V. Heinz, O. Schlüter, and D. Knorr. 2001. Kinetic studies on high-pressure inactivation of *Bacillus stearothermophilus* spores suspended in food matrices. *Innovative Food Science and Emerging Technologies* 2:261–272.

Ardia, A. 2004. Process considerations on the application of high pressure treatment at elevated temperature levels for food preservation. Doctoral thesis, TU Berlin, Fakultät III.

Atrih, A., and S.J. Foster. 2002. Bacterial endospores: The ultimate survivors. *International Dairy Journal* 12:217–223.

Cerf, O. 1977. Tailing of survival curves of bacterial spores. *Journal of Applied Bacteriology* 42:1–19.

Cléry-Barraud, C., A. Gaubert, P. Masson, and D. Vidal. 2004. Combined effects of high hydrostatic pressure and temperature for inactivation of *Bacillus anthracis* spores. *Applied and Environmental Microbiology* 70:635–637.

Collins, M.D., and A.K. East. 1998. Phylogeny and taxonomy of the food-borne pathogen *Clostridium botulinum* and its neurotoxins. *Journal of Applied Microbiology* 84:5–17.

de Heij, W.B.C., L.J.M.M. van Schepdael, R. Moezelaar, H. Hoogland, A.M. Matser, and R.W. van den Berg. 2003. High-pressure sterilization: Maximizing the benefits of adiabatic heating. *Food Technology* 57:37–41.

Gänzle, M.G., H.M. Ulmer, and R.F. Vogel. 2001. High pressure inactivation of *Lactobacillus plantarum* in a model beer system. *Journal of Food Science* 66:1174–1181.

Gola, S., and P.P. Rovere. 2005. Resistance to high hydrostatic pressure of some strains of *Clostridium botulinum* in phosphate buffer. *Industria Conserve* 80:149–157.

Han, Y.W. 1975. Death rates of bacterial spores: Nonlinear survivor curves. *Canadian Journal of Microbiology* 21:1464–1467.

Heinz V., and D. Knorr. 1998. "High pressure germination and inactivation kinetics of bacterial spores." In: *High Pressure Food Science, Bioscience and Chemistry*, ed. N. S. Isaacs, pp. 435–441. Cambridge, UK: Royal Society of Chemistry.

Herdegen, V. 1998. Hochdruckinaktivierung von Mikroorganismen in Lebensmitteln und Lebensmittelreststoffen. Doctoral thesis, TU München, Fakultät für Brauwesen, Lebensmitteltechnologie und Milchwissenschaft, Verlag Ulrich E. Grauer, Stuttgart, Germany.

Igura, N., Y. Kamimura, M.S. Islam, M. Shimoda, and I. Hayakawa. 2003. Effects of minerals on resistance of *Bacillus subtilis* spores to heat and hydrostatic pressure. *Applied and Environmental Microbiology* 69:6307–6310.

Kalchayanand, N., A. Sikes, C.P. Dunne, and B. Ray. 1998. Factors influencing death and injury of foodborne pathogens by hydrostatic pressure pasteurization. *Food Microbiology* 15:207–214.

Kort, R., A.C. O'Brien, I.H.M. van Stokkum, S.J.C.M. Oomes, W. Crielaard, K.J. Hellingwerf, and S. Brul. 2005. Assessment of heat resistance of bacterial spores from food product isolates by fluorescence monitoring of dipicolinic acid release. *Applied and Environmental Microbiology* 71:3556–3564.

Lee, S.-Y., R.H. Dougherty, and D-H. Kang. 2002. Inhibitory effects of high pressure and heat on *Alicyclobacillus acidoterrestris* spores in apple juice. *Applied and Environmental Microbiology* 68:4158–4161.

Malidis, C.G., and J. Scholefield. 1985. The release of dipicolinic acid during heating and its relation to the heat destruction of *Bacillus stearothermopilus* spores. *Journal of Applied Bacteriology* 59:479–486.

Margosch, D. 2004. Behaviour of bacterial endospores and toxins as safety determinants in low acid pressurised food. Doctoral thesis, TU München, Fakultät Wissenschaftszentrum Weihenstephan.

Margosch, D., M.A. Ehrmann, R. Buckow, V. Heinz, R.F. Vogel, and M.G. Gänzle. 2006. High pressure mediated survival of *Clostridium botulinum* and *Bacillus amyloliquefaciens* endospores at high temperature. *Applied and Environmental Microbiology* 72:3476–3481.

Margosch, D., M.A. Ehrmann, M.G. Gänzle, and R.F. Vogel. 2004b. Comparison of pressure and heat resistance of *Clostridium botulinum* and other endospores in mashed carrots. *Journal of Food Protection* 67:2530–2537.

Margosch, D., M.G. Gänzle, M.A. Ehrmann, R.F. Vogel. 2004a. Pressure inactivation of *Bacillus* endospores. *Applied and Environmental Microbiology* 70:7321–7328.

Molina-Gutierrez, A., V. Stippl, A. Delgado, M.G. Gänzle, and R.F. Vogel. 2002. Effect of pH on pressure inactivation and intracellular pH of *Lactococcus lactis* and *Lactobacillus plantarum*. *Applied and Environmental Microbiology*, 68:4399–4406.

Nicholson, W.L., N. Munakata, G. Horneck, H.J. Melosh, and P. Setlow. 2000. Resistance of *Bacillus* endospores to extreme terrestrial and extraterrestrial environments. *Microbiology and Molecular Biology Reviews* 64:548–572.

Owen, B.B., and S.R. Brinkley. 1941. Calculation of the effect of pressure upon ionic equilibria in pure water and in salt solutions. *Chemistry Reviews* 29:461–474.

Paidhungat M., K. Ragkousi, and P. Setlow. 2001. Genetic requirements for induction of germination of spores of *Bacillus subtilis* by Ca^{2+}-dipicolinate. *Journal of Bacteriology* 183:4886–4893.

Paidhungat M., B. Setlow, W.B. Daniels, D. Hoover, E. Papafragkou, and P. Setlow. 2002. Mechanisms of induction of germination of *Bacillus subtilis* spores by high pressure. *Applied and Environmental Microbiology* 68:3172–3175.

Paidhungat, M., B. Setlow, A. Driks, and P. Setlow. 2000. Characterization of spores of *Bacillus subtilis* which lack dipicolinic acid. *Journal of Bacteriology* 182:5505–5512.

Rajan, S., J. Ahn, V.M. Balasubramaniam, and A.E. Yousef. 2006. Combined pressure-thermal inactivation kinetics of *Bacillus amyloliquefaciens* spores in egg patty mince. *Journal of Food Protection* 69:853–860.

Reddy, R., H.M. Solomon, G.A. Fingerhut, E.J. Rhodehamel, V.M. Balasubramaniam, and S. Palaniappan. 1999. Inactivation of *Clostridium botulinum* type E spores by high pressure processing. *Journal of Food Safety* 19:277–288.

Reddy, N.R., H.M. Solomon, R.C. Tetzloff, and E.J. Rhodehamel. 2003. Inactivation of *Clostridium botulinuum* type A spores by high-pressure processing at elevated temperatures. *Journal of Food Protection* 66:1402–1407.

Reddy, N.R., R.C. Tetzloff, H.M. Solomon, and J.W. Larkin. 2006. Inactivation of *Clostridium botulinum* nonproteolytic type B spores by high pressure processing at moderate to elevated high temperatures. *Innovative Food Science and Emerging Technologies* 7:169–175.

Roberts, C.M., and D.G. Hoover. 1996. Sensitivity of *Bacillus coagulans* spores to combinations of high hydrostatic pressure, heat, acidity and nisin. *Journal of Applied Bacteriology* 91:1582–1588.

Rovere, P., S. Gola, A. Maggi, N. Sacamuzza, and L. Miglioli. 1998. "Studies on bacterial spores by combined high pressure-heat treatments: Possibility to sterilize low acid foods." In: *High Pressure Food Science, Bioscience and Chemistry,* ed. N.S. Isaacs, pp. 354–363. Cambridge, UK: Royal Society of Chemistry.

Sale, J.H., G.W. Gould, and W.A. Hamilton. 1970. Inactivation of bacterial spores by hydrostatic pressure. *Journal of General Microbiology* 60:323–334.

San Martín, M.F., G.V. Barbosa-Cánovas, and B.G. Swanson. 2002. Food processing by high hydrostatic pressure. *Critical Reviews in Food Science and Nutrition* 42:627–645.

Sizer, C.E., V.M. Balasubramaniam, and E. Ting. 2002. Validating high-pressure processes for low-acid foods. *Food Technology* 56:36–42.

Smeller, L. 2002. Pressure-temperature phase diagrams of biomolecules. *Biochimica Biophysica Acta* 1595:11–29.

Smelt, J.P.P.M., J.C. Hellemons, P.C. Wouters, and S.J.C. van Gerwen. 2002. Physiological and mathematical aspects in setting criteria for decontamination of foods by physical means. *International Journal of Food Microbiology* 78:57–77.

Stewart, C.M., C.P. Dunne, A. Sikes, and D.G. Hoover. 2000. Sensitivity of spores of *Bacillus subtilis* and *Clostridium sporogenes* PA 3679 to combinations of high hydrostatic pressure and other processing parameters. *Innovative Food Science & Emerging Technologies* 1:49–56.

Stringer, S.C., M.D. Webb, S.M. George, C. Pin, and M.W. Peck. 2005. Heterogeneity of times required for germination and outgrowth from single spores of nonproteolytic *Clostridium botulinum*. *Applied and Environmental Microbiology* 71:4998–5003.

Wuytack, E.Y., S. Boven, and C.W. Michiels. 1998. Comparative study of pressure-induced germination of spores at low and high pressures. *Applied and Environmental Microbiology* 64:3220–3224.

Wuytack, E.Y., and C.W. Michiels. 2001. A study on the effects of high pressure and heat on *Bacillus subtilis* spores at low pH. *International Journal of Food Microbiology* 64:333–341.

Wuytack, E.Y., J. Soons, F. Poschert, and C.W. Michiels. 2000. Comparative study of pressure- and nutrient-induced germination of *Bacillus subtilis* spores. *Applied and Environmental Microbiology* 66:257–261.

Chapter 6

The Quasi-chemical and Weibull Distribution Models of Nonlinear Inactivation Kinetics of *Escherichia coli* ATCC 11229 by High Pressure Processing

Christopher J. Doona, Florence E. Feeherry, Edward W. Ross, Maria Corradini, and Micha Peleg

Introduction

High pressure processing (HPP) is an emerging food processing technology and one of the leading alternatives to replacing thermal food processing in the drive to meet increasing consumer demand for foods featuring improved organoleptic qualities and higher acceptance. In general, high pressure treated foods feature more fresh-like character, higher nutrient retention, and improved sensory attributes such as flavor, color, and texture than their thermally treated counterparts. For HPP food products to be successful in the marketplace, they must appeal to the consumer and they must be free of microbiological spoilage organisms and safe from infectious pathogens and bacterial spores.

Predictive mathematical models provide convenient tools for assessing microbial behavior in foods and ensuring the microbiological safety and stability of foods (McMeekin et al., 1993). Depending on the needs of a particular application, one can choose from a variety of existing mathematical models to characterize microbial kinetics in foods (Baranyi and Roberts, 2000; Membré et al., 1997; McMeekin et al., 1997; Whiting et al., 1996; Whiting, 1995; Skinner et al., 1994; Whiting and Buchanan, 1994; Buchanan et al., 1993; McMeekin et al., 1993; Whiting, 1993; and Buchanan, 1992). Empirical models based on sigmoidal equations (e.g., logistic equation, Gompertz function, Baranyi-Roberts model) are generally used to describe either microbial growth or inactivation kinetics (Peleg, 2003; Legan et al., 2002; Baranyi and Roberts, 2000; Baranyi and Roberts, 1997; McMeekin et al., 1997; Whiting, 1995; Kamau et al., 1990). Traditionally, the log linear (aka, first-order kinetics) model has been used to describe the inactivation of microbes in thermally processed foods.

Lethality (denoted in terms of an F_0 value) is the benchmark used by the food industry to measure the safety of thermal processes. The assumption inherent in the use of the F_0 value is that isothermal survival curves in the lethal regime follow first-order kinetics. The reciprocal of the slope of the log linear survival ratio versus time defines the "D-value." The temperature dependence of D is assumed to be log linear with a slope characterizing the quantity z. The D-value at a reference temperature and the z-value characterize the heat resistance of a target organism. Knowing the D-z parameters for a target organism allows the calculation of the lethality for any temperature-time process in the lethal regime in terms of an F_0 value that can be represented as an equivalent processing time at a select reference temperature. There is a growing body of experimental results showing that isothermal microbial survival curves of bacterial cells and spores are not strictly log linear (van Boekel, 2002; Peleg and Cole, 1998), and alternative approaches to modeling inactivation curves might be needed in such circumstances.

The survival curves for organisms treated with HPP do not completely adhere to log linear kinetics (e.g., Ahn et al., 2007; Chen, 2007; Guan et al., 2006; Margosch et al., 2006; Chen and Hoover, 2004; Margosch et al., 2004a, 2004b). The inactivation kinetics in these cases can exhibit nonlinearities such as "shoulders" (Doona et al., 2005; Feeherry et al., 2005; Ross et al., 2005) or "tailing" (Tay et al., 2003). The first-order kinetics model does not inherently account for these nonlinearities. Consequently, the safe preservation of foods by HPP requires the development of alternative kinetics models that can account for the nonlinear features of the survival curves and accurately predict the inactivation patterns observed with harmful organisms (Heldman and Newsome, 2003).

This chapter presents two alternative modeling approaches for characterizing nonlinear inactivation kinetics called the Quasi-chemical kinetics model (Taub et al., 2003; Feeherry et al., 2003, 2001) and the Weibull distribution model. The Quasi-chemical model was developed (Taub et al., 2003) to control foodborne pathogens in intermediate moisture (IM) foods using combinations of "hurdles." The model was developed as part of an effort by the US Army Natick Soldier Research, Development, & Engineering Center to develop bi-layered, savory sandwiches (barbecue chicken or nacho cheese and beef sausage filling) that feature high consumer acceptance ratings. These pocket sandwiches have received extensive coverage in the media (ABC News, 2002; BBC News, 2002; CNN.com, 2002; Cook, 2002; Fabricant, 2002; Graham-Rowe, 2002; Yahoo! News, 2002). The Quasi-chemical model belongs to a distinct group of predictive microbial models (Del Nobile et al., 2003; Membré et al., 1997; Peleg, 1996; Jones et al., 1994; Jones and Walker, 1993; Whiting and Cygnarowicz-Provost, 1992) that are capable of characterizing continuous growth-death kinetics observed with appropriate combinations of hurdles (Figure 6.1).

Based on its ability to describe nonlinear microbial inactivation kinetics occurring through combinations of hurdles (Figure 6.1), the Quasi-chemical model is a suitable candidate for modeling inactivation kinetics with HPP using the same mathematical framework (Doona et al., 2005; Feeherry et al., 2005; Ross et al., 2005). The Weibullian model derives from the notion that survival curves are the cumulative form of the temporal distribution of mortality events. Accordingly, the logarithmic inactivation rate can be time dependent, even under isothermal and isobaric conditions. The Weibull distribution equation is a versatile and flexible model that can describe log linear and non-log linear kinetics. These survival curves can exhibit a shoulder and various degrees of upward or downward concavity. It should also be noted that according to the Weibullian model, a "shoulder" is a manifestation of a standard deviation that is considerably smaller than the mean or mode (Peleg, 2006). The Weibull distribution model has been effective in describing the isothermal nonlinear inactivation kinetics of pathogens such as *Listeria monocytogenes* (Chen and Hoover, 2004) and a large number of bacteria and certain clostridia and bacilli spores (Peleg, 2006; van Boekel, 2002).

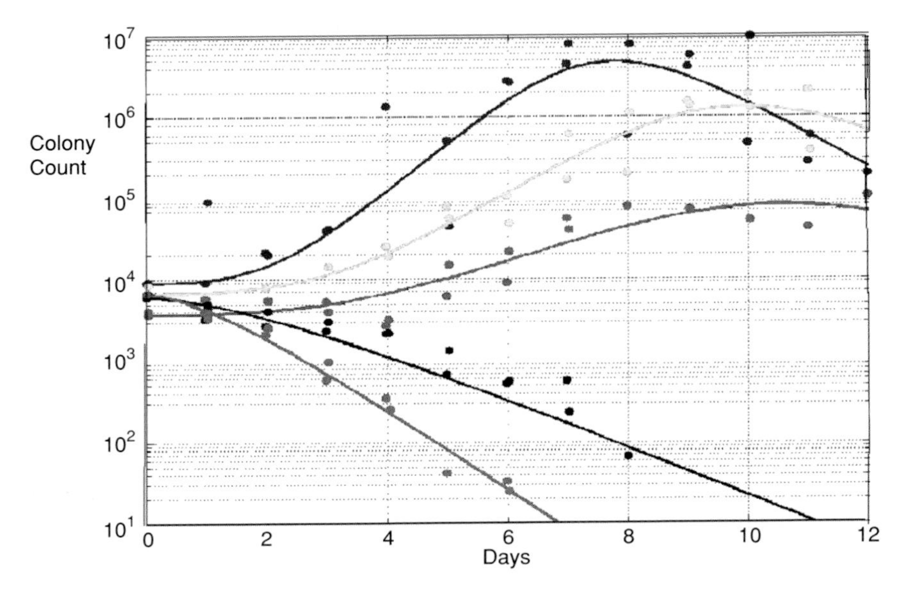

Figure 6.1. Quasi-chemical modeling (smooth curves) of growth-death and death-only kinetics (solid symbols) data for *Staphylococcus aureus* in bread over the pH range of 5.4–4.9 (5.38, 5.36, 5.31, 5.19, and 4.97, in descending order of the depicted curves) and $a_w = 0.86$ and $T = 35°C$.

This chapter demonstrates the application of each of these models to the same sets of *Escherichia coli* ATTC 11229 survival data obtained with HPP treatments at various high pressure and temperature conditions. Despite the differences inherent in the Quasi-chemical and Weibull distribution models, each model works effectively in capturing the phenomenon of microbial inactivation. The dependence of each model's parameters on high pressure and temperature can be used to create predictive secondary models. The equivalence chart is derived from the Quasi-chemical model and can predict a range of combinations of high pressure, temperature, and processing times that inactivate 10^6 target pathogens. The equivalence chart can be used to guide food technologists in the selection of appropriate processing conditions to ensure the microbial safety of HPP food products (Feeherry et al., 2005). Secondary models are also built from the high pressure and temperature dependences of the Weibullian model parameters, and they could, in principle, be used to accomplish a similar result (Peleg, 2006). We commend to readers interested in using the Quasi-chemical model the

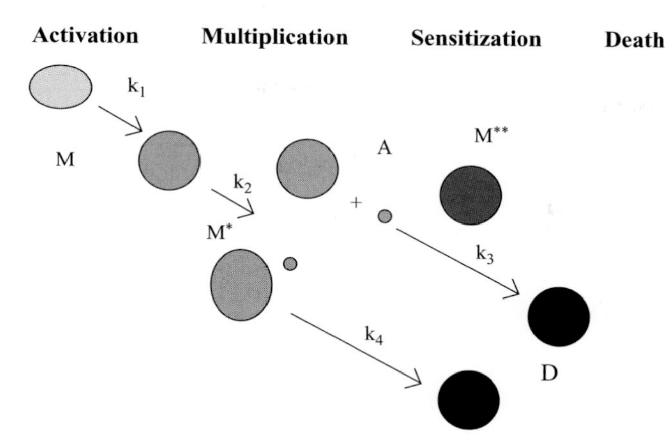

Figure 6.2. Schematic mechanism of the Quasi-chemical model denoting cells in the stages of metabolizing (M), multiplying (M*), sensitization to death (M**), and dead (D), and the hypothetical antagonist (A).

software code, a worked example, and more a rigorous mathematical discussion that was published previously (Ross et al., 2005, Appendices A and B).

Theoretical Background

Quasi-Chemical Model

The microbial lifecycle consists of four phases (lag, exponential, stationary, and death). When modeling microbial behaviors in foods, the Quasi-chemical model accounts for all four phases of the microbial lifecycle contiguously (as shown in Figure 6.1). The theoretical basis for the Quasi-chemical model is a hypothetical mechanism (Figure 6.2) that describes the events occurring at the sub-cellular level as a sequence of steps, much like a chemical reaction (Ross et al., 2005; Taub et al., 2003; Feeherry et al., 2001).

Any mechanism that attempts to link the macroscopic growth-death kinetics of a microbial population with events at the cellular and molecular levels requires drastic simplifications to describe the network of myriad biochemical processes occurring in the microorganisms (Bray and White, 1966; Hinshelwood, 1953). Simplified mechanisms such as these can furnish insight into complex biochemical processes. A similar

Table 6.1. Summary of mechanism and equations for the Quasi-chemical model

Reaction Step	Rate Equations	ODEs	
$M \rightarrow M^*$ (activation)	$v_1 = k_1M$	$dM/dt = (-v_1) = -k_1M$	Eq. (1)
$M^* \rightarrow 2M^* + A$ (multiplication)	$v_2 = k_2M^*$	$dM^*/dt = (v_1 + v_2 - v_3 - v_4) = k_1M + M^*(G - \varepsilon A)^{\ddagger}$	Eq. (2)
$M^* + A \rightarrow D$ (sensitization)	$v_3 = k_3M^*A$	$dA/dt = (v_2 - v_3) = M^*(k_2 - \varepsilon A)^{\ddagger}$	Eq. (3)
$M^* \rightarrow D$ (natural death)	$v_4 = k_4M^*$	$dD/dt = (v_3 - v_4) = M^*(k_4 + \varepsilon A)^{\ddagger}$	Eq. (4)
		$U(t) = M(t) + M^*(t)$	Eq. (5)

‡ E is a scale factor with $E = 1 \times 10^{-9}$

type of approach has proven effective for understanding nonlinear dynamics in chemical and biological systems (Epstein and Pojman, 1998), such as slime mold aggregation (Goldbeter, 1991) and the prey-predator cycle of planktonic rotifers feeding on algae in a chemostat (Fussman et al., 2000).

The mechanism postulated for the Quasi-chemical model involves autocatalytic growth and negative feedback (Figure 6.2). Presumably, the latter is due to the formation of an antagonistic, diffusible metabolite as an intermediate that acts as an intercellular signaling molecule in the process known as quorum sensing (Smith et al., 2004; Bassler, 1999; Dunny and Winans, 1999; England et al., 1999; Novick, 1999; Kleerebezem et al., 1997; Goldbeter, 1996). The steps of the mechanism are also listed in Table 6.1.

The first step of the mechanism is called "activation" (Figure 6.2) and represents the conversion of metabolizing cells in the lag phase to metabolizing cells capable of dividing ($M \rightarrow M^*$ with rate constant k_1). The second step depicts the multiplication of cells via binary division with the concomitant formation of an antagonistic metabolite as an intermediate ($M^* \rightarrow 2M^* + A$ with the rate constant k_2). The third and fourth steps represent alternative termination pathways. In the third step, A interacts with the bacterium and sensitizes it in a process that accelerates its death through the intermediate M** (A + M* → M** → D and characterized by the rate constant k_3). The fourth step depicts termination of the cells through "natural death" (M* → D with the rate constant k_4).

From this hypothetical mechanism, one can construct a corresponding set of kinetics rate equations with associated rate constants and a system of ordinary differential equations (ODEs) that constitute the basis of the Quasi-chemical model (Table 6.1). The rate equations interrelate the velocities (v) of each step as the product of the rate constant (k) and the "concentrations" of the participating entities, in a manner similar to chemical rate equations (Fussman et al., 2000). The rate constants are constrained to only non-negative values. The system of ODEs (equations 6.1–6.4 in Table 6.1) reflects the processes responsible for the formation and/or elimination of the entities M, M*, A, and D. The quantity U represents the total microbial plate counts without distinguishing lag-phase and growth-phase cells:

$$\mathrm{U} = \mathrm{M} + \mathrm{M}^* \text{ (in units of CFU/mL)} \tag{6.5}$$

The Weibull Model (Empirical Approach)

Let us assume that the cumulative form of the Weibull distribution function (equation 6.6) or the empirical power law model (Peleg and Cole, 1998; van Boekel, 2002; Peleg, 2003; Peleg, 2006) can characterize the semi-logarithmic survival curve for *E. coli* under ideal isothermal and isobaric conditions according to

$$\log_{10} S(t) = -b(P,T)t^{n(P,T)} \tag{6.6}$$

in which the variables N(t) and N_o in the survival ratio S(t) = N(t)/N_o are the cells' momentary ("instantaneous") and initial numbers, respectively, and b(P,T) and n(P,T) are pressure- and temperature-dependent coefficients. When n(P,T) ≠ 1 in equation 6.6 (Peleg and Cole, 1998), the isothermal and isobaric inactivation rate is always time dependent (equation 6.7).

$$\left|\frac{d\log_{10} S(t)}{dt}\right|_{\substack{P=const\\T=const}} = -b(P,T)n(P,T)T^{n(P,T)-1} \tag{6.7}$$

The coefficient b(P,T) can be considered as a "rate parameter" of sorts. The exponent n(P,T) is a measure of the concavity of the isothermal and isobaric semi-logarithmic survival curve. According to this model, values of n(P,T) > 1 imply that the semi-logarithmic survival curve has downward concavity. Values of n(P,T) < 1 indicate the survival curve has upward concavity, and n(P,T) = 1 indicates a straight line. Downward concavity in the survival curve, the condition with n(P,T) > 1, may indicate that damage accumulating in the survivors sensitizes them to further injury or inactivation. In contrast, upward concavity, n(P,T) < 1, suggests that the sensitive members of the population die quickly, leaving progressively more resistant survivors, as is seen with "tailing." The destruction of these more resistant survivors requires increasing the duration of their exposure to the HPP conditions (Peleg and Penchina, 2000). A log linear survival curve, n(P,T) = 1, implies that the probability of death is time independent, which is an unlikely scenario, or that the existence of subpopulations of the former kinds balance one another exactly, which is also improbable.

Ad hoc empirical expressions can be used to describe the pressure and temperature dependence of the Weibullian-power law model's parameters. For a given organism, the pressure effect is felt only above a certain threshold value, and the temperature effect is also felt only once it has reached a certain level. The log-logistic model can account for these types of responses (Peleg, 2006):

$$b(P)|_{T=const} = \log_e\{1 + \exp\{-k_P(T)[P - P_c(T)]\} \tag{6.8}$$

or

$$b(T)|_{P=cont} = \log_e[1 + \exp\{-k_T(P)[T - T_c(P)]\} \tag{6.9}$$

in which $k_P(T)$ and $P_c(T)$ are temperature-dependent coefficients and $k_T(P)$ and $T_c(P)$ are pressure-dependent coefficients. The pressure and temperature dependences of n(P,T) are even qualitatively less predictable, but they can be derived directly from the survival curves themselves (see below). Representative examples (Peleg, 2006) can be expressed as

$$n(P)|_{T=const} = c_{0P}(T)\exp[-c_{1P}(T)P] \tag{6.10}$$

in which $c_{0P}(T)$ and $c_{1P}(T)$ are temperature-dependent coefficients, and

$$n(T)|_{P=const} = c_{0T}(P)\exp[-c_{1T}(P)T] \tag{6.11}$$

in which $c_{0T}(P)$ and $c_{1T}(P)$ are pressure-dependent coefficients.

According to this modeling approach, accurate characterization of an organism's resistance or sensitivity to the effects of an HPP treatment that combines high pressure and temperature may require at least four survival parameters. In the above example, the parameter set could be $\{k_p, P_c, c_{0P}, \text{and } c_{1P}\}$ or $\{k_t, T_c, c_{0T}, \text{and } c_{1T}\}$. The number of parameters could be reduced if n(P,T) could be approximated by a single representative value. Values for these parameters can, in principle, be determined experimentally from survival curves obtained at various constant high pressure-temperature combinations. Once these constants are determined, they can be used to determine the organism's isothermal-isobaric survival curve under any pressure-temperature combination covered within the range of the experimental data used to determine them.

It can be demonstrated that the Weibullian-Power Law model (equation 6.6) is not unique and several plausible alternative mathematical expressions could replace it (Peleg and Penchina, 2000). As long as these alternative models yield comparable fits to the experimental survival data, and the temperature and pressure dependence of their coefficients can be expressed algebraically, these alternatives could be used interchangeably (Peleg, 2003).

Demonstration of the Two Modeling Approaches

Inactivation (death-only) kinetics data for the high pressure inactivation of *E. coli* ATCC 11229 in the aqueous whey protein Surrogate Food System (SFS) were collected at systematically varied increments of pressure and temperature (Doona et al., 2005; Feeherry et al, 2005; Ross et al., 2005). Detailed technical information regarding the experimental protocol used in these studies is available from Feeherry et al. (2005) and discussed only briefly here. HPP experiments were carried out in a 2-liter EPSI high pressure unit with temperature control. Denaturation of the whey protein occurred in some processing conditions, but the conversion of the aqueous protein suspension to a gel did not alter the form of the observed inactivation kinetics or inhibit the ability to recover cells in these experiments. Inactivation plots were constructed from colony counts obtained using standard spread plating and enumeration techniques. In the range of conditions examined, the inactivation curves tended to exhibit a shoulder at lower pressures (P = 207 MPa, 30 kpsi) and temperatures (T = 30°C), and the kinetics curves became decidedly linear with increases in pressure (P = 370 MPa, 53.7 kpsi) and temperature (T = 50°C). In all cases, the Quasi-chemical model and the Weibullian model yielded fits in good agreement with the inactivation kinetics data, based on statistical criteria.

Primary Models

The Quasi-chemical Model

The Quasi-chemical model was fit to the data for each of the kinetics profiles with nonlinear least-squares regression analysis (Figure 6.3A and B).

In general, the least-squares fitting procedure worked well, since the data were not particularly noisy. Fitting the data was not entirely mechanical and the goodness of fit depended on making fortuitous initial estimates of k_1 through k_4, then solving the ODEs to determine calculated values of the colony counts (Ross et al., 2005). The calculated value was compared to the plate count data, and the rate constant values (k_1 through k_4) were changed iteratively to minimize the discrepancy between the calculated and experimental values according to nonlinear least-squares procedures. The value of the inactivation rate constant (μ) was estimated by numerically differentiating

$$L(t) = \log_{10}(U(t)) \tag{6.12}$$

and determining the minimum slope of L(t). If the time at which μ occurs is defined as t_μ, then the lag (denoted λ) is estimated from the formula

$$\lambda = t_\mu - (L_\mu - L_o)/\mu \tag{6.13}$$

in which $L_o = L(0)$ and $L_\mu = L(t_\mu)$.

It should be noted that in these processing conditions, the relatively slow inactivation of *E. coli* required in some cases data acquisition up to 5 hr, which is beyond a practicable timeframe for food processing in industry. It is possible that longer data acquisition times could reveal more detail in the kinetics profiles (such as tailing) that would change the appearance of the survival curve. It is unlikely that collecting more

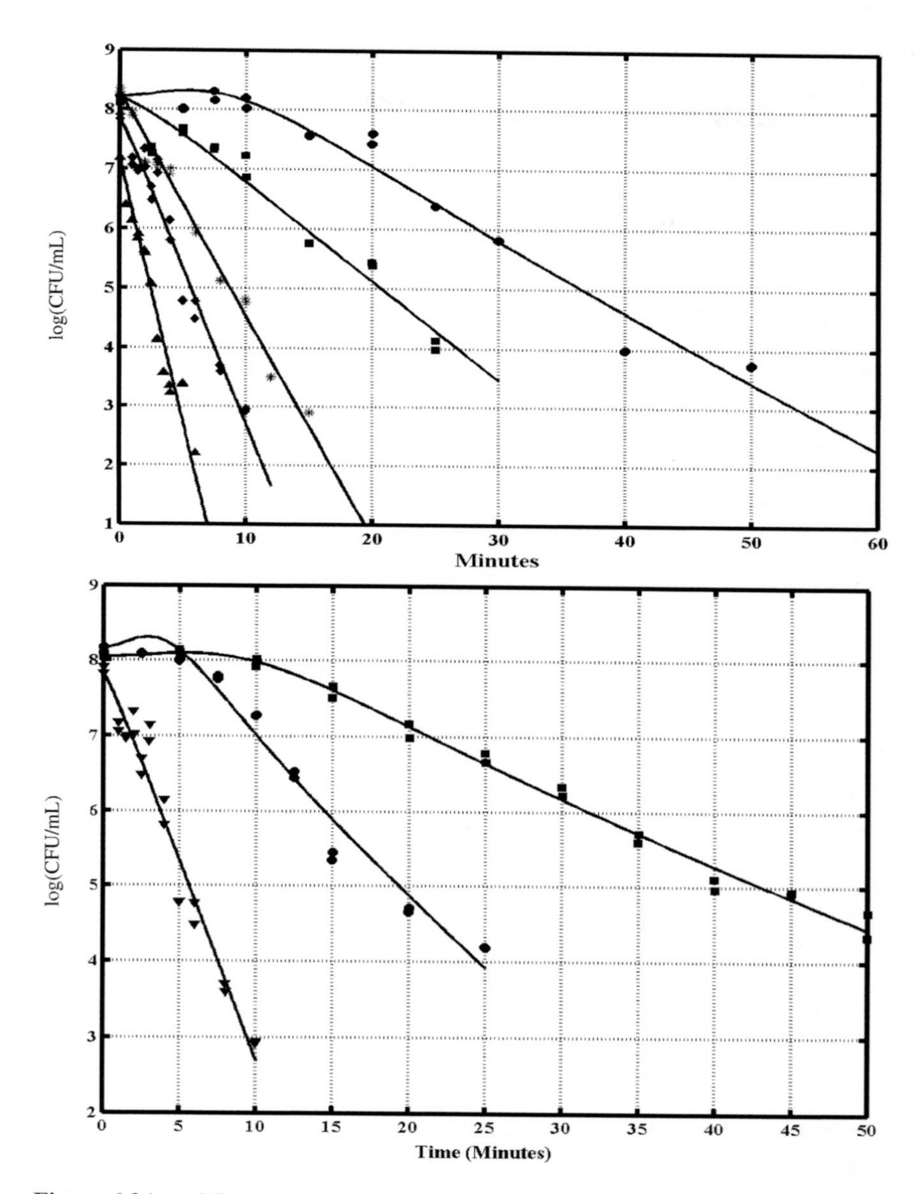

Figure 6.3A and B. Inactivation curves of *E. coli* data (solid symbols) as functions of (A) pressure = 30, 38.7, 43.7, 48.7, 53.7 kpsi (in descending order) at 50°C, and of (B) temperature = 30, 40, 50°C (in descending order) at 48.7 kpsi.

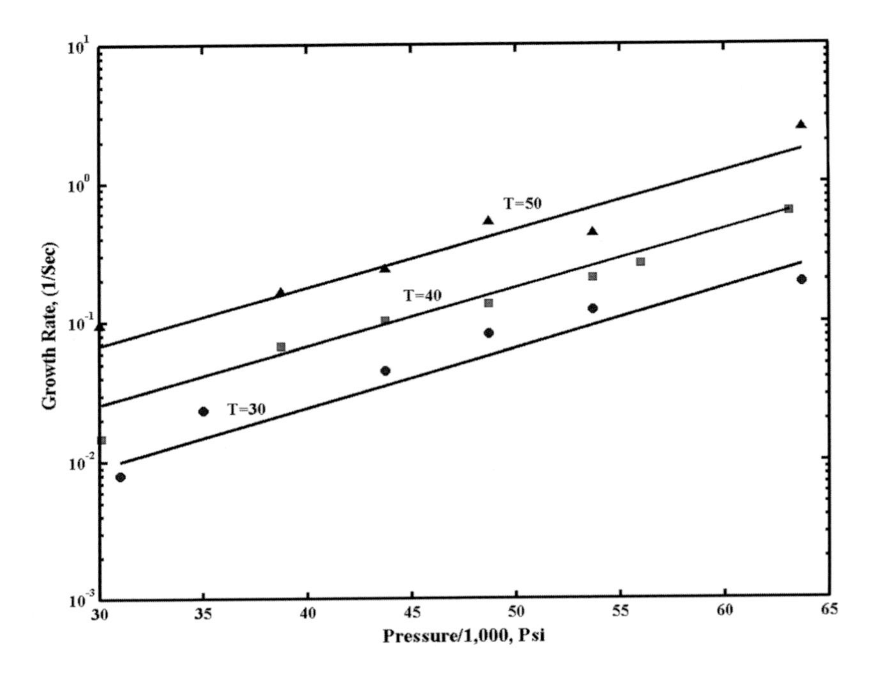

Figure 6.4. Log-linear relationship of the inactivation rate constants calculated estimated with the Quasi-chemical model at various high pressures and T = 50°C (▲), T = 40°C (■), and T = 30°C (•).

data in this region of the survival curves would significantly impact the estimates of the inactivation rate constants (μ).

The log (μ) versus pressure was a linear relationship (Figure 6.4), similar to what one might find with the pressure-dependence of chemical reaction rate constants (Fábián and van Eldik, 1993). The estimated values of μ depicted in Figure 6.4 reflect the HPP inactivation of *E. coli* when the supporting menstrum is whey protein. The properties of the suspension medium influence the observed kinetics and can change the estimated value of μ accordingly. Figure 6.5 shows a comparison of the influence of three different substrates on the survival curve for the inactivation of *E. coli* at P = 48.7 kpsi and T = 40°C for whey (top curve), potato starch (middle), or isotonic phosphate buffer (bottom). On a relative scale, whey protein appears to exert a discernible protective effect, buffer solution is least protective, and the influence of starch is

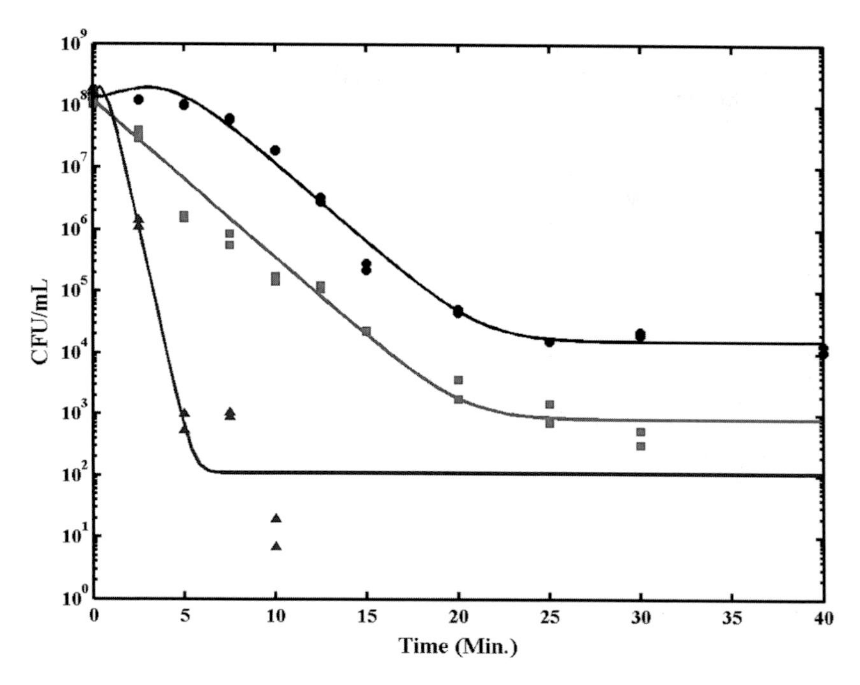

Figure 6.5. The nature of the food matrix affects the rate of *E. coli* inactivation at P = 45 kpsi and T = 40°C. Whey protein (top curve) exerts the strongest protective effect, followed by starch (middle curve), and isotonic phosphate buffer (bottom curve).

intermediate that of whey and buffer. The effects of HPP on a starch SFS and its retrogradation behavior have been reported previously (Doona et al., 2006).

The Weibullian-Power Law

The Weibullian-Power Law (equation 6.6) is also an effective primary model for describing the *E. coli* survival curves. The fits of the data are shown in Figure 6.6A, B, and C. The model could be judged as adequate by statistical criteria (Table 6.2). The figures and tables show a concavity inversion in the inactivation curves. In the lower pressure regimes at all three temperatures, the survival curves tend to show downward concavity consistent with $n(P,T) > 1$. As the pressure increases, the survivor curves become progressively more linear, which is manifested in the relation $n(P,T) = 1$. At the highest pressure regions of the

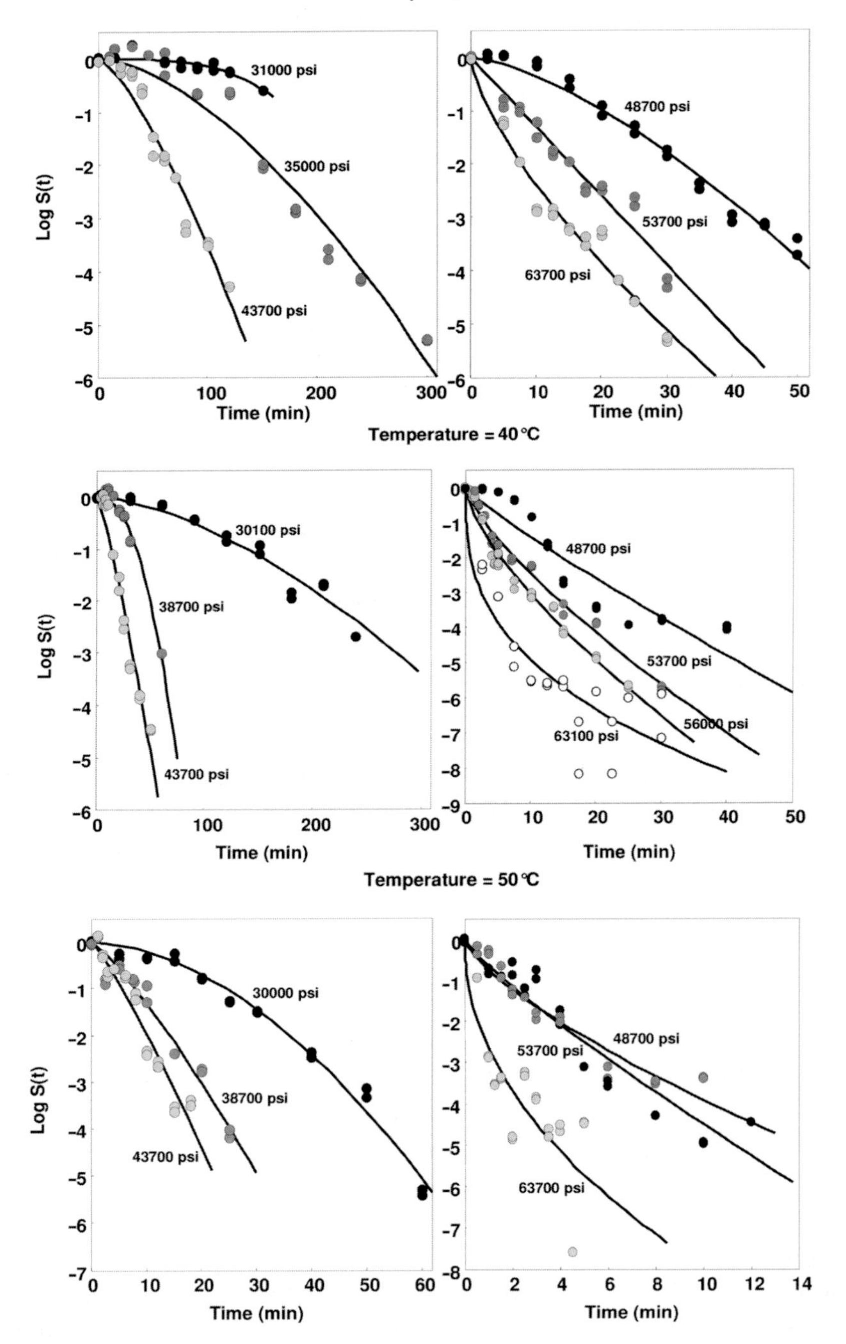

Figure 6.6A, B and C. The survival curves of *E. coli* during three "isothermal" processes, fitted with the Weibull distribution (equation 6.1) primary model.

Table 6.2. Survival parameters of *E. coli* subjected to HPP treatments at three different temperatures

	n(P)[1]			b(P)[2]		
Temperature (°C)	c_1	c_2	MSE (−)	k_P (psi^{-1})	P_c (psi)	MSE (−)
30	13.3	0.000049	0.222	0.00018	66431	0.0005
40	41.7	0.000077	0.030	0.00029	56079	0.002
50	5.7	0.000039	0.004	0.00021	51743	0.004

[1] Calculated from equation 10 (exponential) as an ad hoc empirical model.
[2] Calculated from the log logistic model (equation 8).

of the experimental conditions, the mortality rate increases dramatically, and the survival curves show upward concavity, $n(P, T) < 1$, with the implication of resistant subpopulations showing "tailing."

Secondary Models

Secondary models describe the interrelation of the primary model's parameters with, in the case of HPP, the processing parameters of high pressure (P) and temperature (T). Specifically, secondary models will estimate the pressure and temperature dependences of μ and λ model for the Quasi-chemical model, and n and b for the Weibullian-Power Law model.

The Quasi-Chemical Model

The Quasi-chemical model produced results that could be used to generate a unique secondary model designated as an equivalence chart that predicts the HPP conditions required to affect a pasteurization process (kill 10^6 *E. coli*) according to the following equation:

$$t_p = \lambda + 6/\mu \tag{6.14}$$

in which t_p is the processing time, λ is the lag time, and μ is the inactivation rate constant.

The relationship of the inactivation rate constants (μ) with the processing parameters high pressure and temperature can be expressed

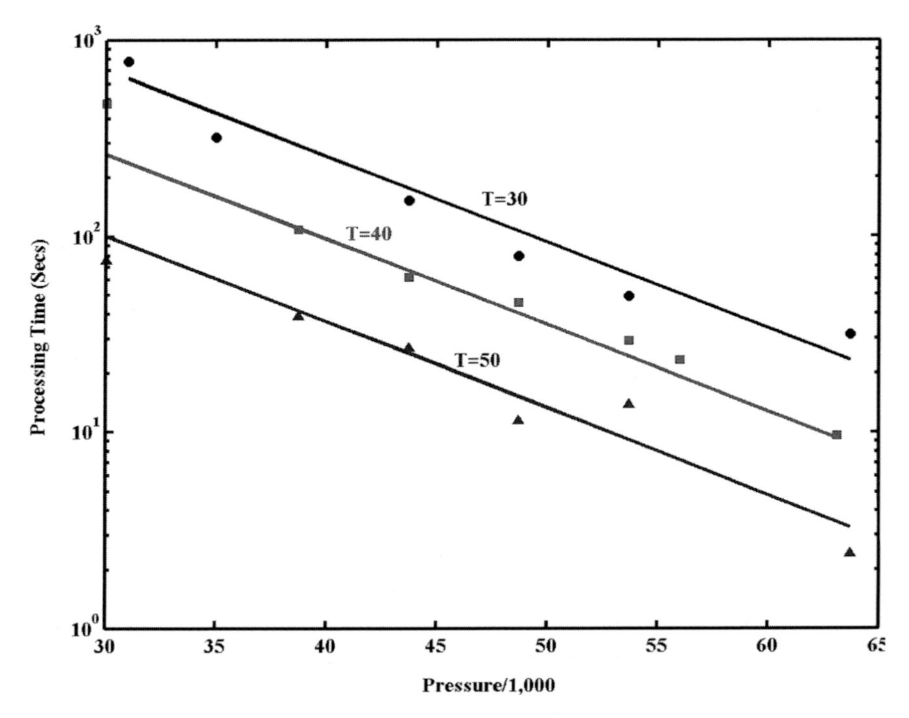

Figure 6.7. Equivalence chart is defined as a secondary model predicting equivalent HPP conditions for inactivating 10^6 *E. coli.* The straight lines represent analysis of the data (•) for T = 30°C, (■) for T = 40°C, and (▲) for T = 50°C.

according to the following linear equation:

$$\log_{10}(\mu) = C_0 + C_T T + C_P P/1000 \tag{6.15}$$

This relationship produced a reasonably accurate fit of the estimated values of μ with P and T ($R^2 = 95.6\%$) and accurate values of the coefficients ($C_0 \approx 4.496 \pm .2007$, $C_T \approx -.0416 \pm .0038$, and $C_P \approx -.0417 \pm .0028$). Combining these results with the experimentally determined lag times produced the resultant equivalence chart (Figure 6.7).

The equivalence chart predicts equivalent combinations of conditions of pressure, temperature, and time that could equally well affect the killing of 10^6 *E. coli* in the present surrogate food system. Several points in the equivalence chart were tested to validate the killing of 10^6 *E. coli* using HPP conditions not used in the construction of the model, and in all cases the results were consistent with the model's predictions.

Such an equivalence chart can guide food product developers in the pasteurization of foods that are sensitive to temperature or pressure. For example, temperature-sensitive foods such as fruit juices can be processed at lower temperatures at higher pressures and/or for longer times. Similarly, pressure-sensitive foods such as fish fillets can be processed at lower pressures and high temperatures or longer times. In both cases, an equivalent HPP treatment can be sought that will ensure an effective pasteurization and food safety without compromising the quality attributes of the product. To get the full benefit of this approach, the equivalence chart should be constructed for individual foods or types of food products.

The Weibullian-Power Law

The Weibullian-Power Law produced results that could be used to generate the following secondary models. Specifically, the log-logistic model (equation 6.8) can describe the pressure dependence of the Weibullian parameter b(P), and an ad hoc empirical model (equation 6.10) can describe the pressure dependence of n(P), as shown in Figure 6.8A, B, and C. The regression parameters of these secondary models are listed in Table 6.3.

The reader should note that the Weibullian model has been constructed without any assumed kinetics order and that the sole source of information for the analysis was the experimentally determined response of the organism to the HPP treatments. It is likely that the magnitude of the organism's response would have been different if the substrate had been an alternative medium. In any case, one might expect that the inactivation kinetics of an organism will follow similar patterns under different permutations of time-temperature-pressure conditions.

Mathematical Properties of the Quasi-Chemical Model

Fitting the Quasi-chemical model to the death-only kinetics observed with HPP yields an estimate of μ (Doona et al., 2005; Feeherry et al., 2005; Ross et al., 2005) and two different sets of values for the rate constant parameters k_1–k_4 (Ross et al., 2005). In particular, the Quasi-chemical model calculates a "fast set" and a "slow set" of values for the rate constant parameters k_1–k_4 (Ross et al., 2005). Both the "fast

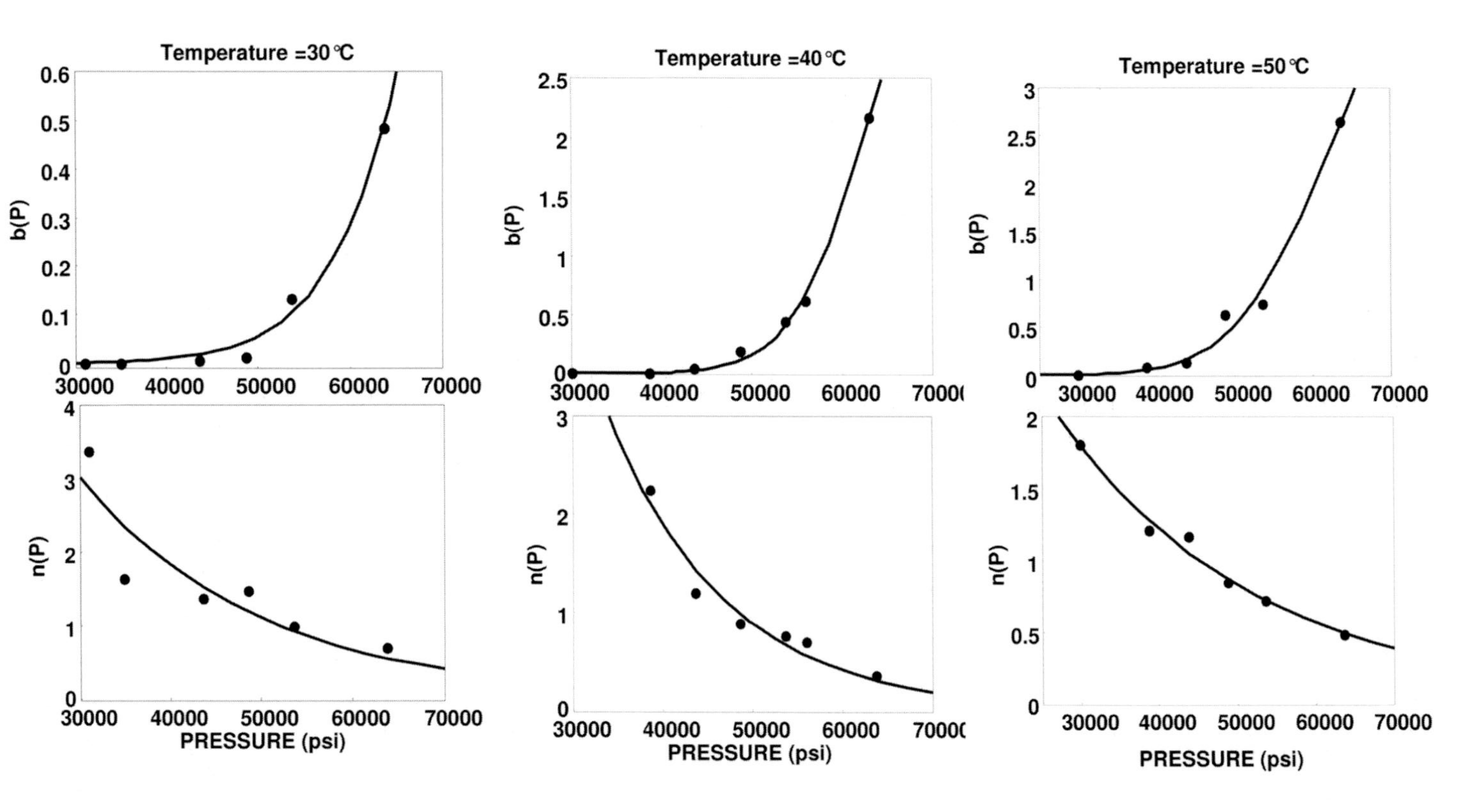

Figure 6.8A, B and C. Secondary models relating the pressure dependences of the Weibullian-Power Law model's parameters at the three processing temperatures by fitting the log-logistic model (equation 6.8) to the pressure dependence of b(P) and an ad hoc empirical model (equation 6.10) to that of n(P).

Table 6.3. Inactivation and regression parameters of *E. coli* subjected to high pressure treatments at three different temperatures fitted with the Weibullian model

Temperature (°C)	Pressure (psi)	b	n	MSE (–)
30	31000	2.7×10^{-8}	3.37	0.003
	35000	0.0005	1.63	0.138
	43700	0.006	1.37	0.095
	48700	0.012	1.47	0.036
	53700	0.132	0.99	0.073
	63700	0.482	0.70	0.069
40	30100	0.0004	1.58	0.036
	38700	0.0003	2.24	0.017
	43700	0.047	1.19	0.130
	48700	0.184	0.89	0.382
	53700	0.438	0.75	0.054
	56000	0.621	0.69	0.085
	63700	2.161	0.36	0.676
50	30000	0.003	1.81	0.046
	38700	0.081	1.21	0.089
	43700	0.135	1.16	0.098
	48700	0.638	0.85	0.235
	53700	0.749	0.72	0.136
	63700	2.653	0.48	0.932

Table 6.4. "Fast set" and "slow set" from fitting microbial inactivation data with the Quasi-chemical model

	k_1	k_2	k_3	k_4	μ
"Fast"	34.30	0.96	0	1.25	−0.12
"Slow"	0.282	0.70	0	19.31	−0.12

set" and the "slow set" of parameters arrive at the same estimate of μ (Table 6.4), have negligible values for k_3, and fit the data equally well (the fitted curves calculated by the "fast" and "slow" sets are nearly identical). One important distinction is that the "fast set" features large values of k_1 ($k_1 \gg 1$, and $M^* \gg M$ shortly after $t = 0$), while the "slow set" has small values of k_1 ($k_1 < 1$, and $M^* \ll M$).

A possible explanation clarifying why the Quasi-chemical model lacks uniqueness and yields two solutions for the rate parameters (the "fast" and "slow" sets) for death-only kinetics is discussed below, and may also be valuable in other circumstances. The differential equations for M and M* (Table 6.1) are re-written as

$$dM/dt = -k_1 M \tag{6.1}$$

$$dM^*/dt - QM^* = k_1 M \tag{6.2}$$

in which $Q = (k_2 - k_3 A - k_4$; see Table 6.1).

In the latter stages of the process, the quantity A becomes constant, which suggests that we examine the behavior of solutions to these equations in that case. The solution of equation 6.1 is known:

$$M = Ie^{-k_1 t} \text{ and } I = M(0) \tag{6.16}$$

The general solution of equation 6.2 when A is constant, and so Q is constant, is

$$M^* = Ce^{Qt} - Ik_1 Y^{-1} e^{-k_1 t} \tag{6.17}$$

in which C is an arbitrary constant of integration, and

$$Y = k_1 + Q. \tag{6.18}$$

Then the plate count U can be written as equation 6.19:

$$U = M + M^* = Ce^{Qt} + IWe^{-k_1 t} \tag{6.19}$$

with $W = Q/Y = Q/(k_1 + Q)$. Equations 6.17 and 6.19 are a general solution to equations 6.1 and 6.2 when A is constant, except when $Y = 0$, a case to be discussed later. Equation 6.19 predicts that U dies away as a weighted sum of two functions with exponential rates (k_1 and $-Q$).

We now consider two extreme cases. First, suppose that

$$Q \ll -k_1 \quad \text{or} \quad Y \ll 0$$

so that, as t increases, the first term in equation 6.19 dies away much faster than the second term. Then equation 6.19 becomes

$$\begin{aligned} U &\approx IWe^{-k_1 t} = WM \\ \log(U) &\approx \log(IW) - (k_1/B)t \\ d(\log(U))/dt &\approx -k_1/B \end{aligned} \tag{6.20}$$

with $B = \log_e 10$. These conditions describe what is called the "slow" case and is characterized by small values of k_1 (Ross et al., 2005). In this situation, U is approximately a multiple of M. Also, U is affected only by k_1, and U is unaffected by changes in Q, as long as $Q \ll -k_1$.

Examining the opposite extreme case when:

$$Q \gg -k_1 \quad \text{or} \quad Y \gg 0$$

so that the second term in equation 6.19 dies away much faster than the first. In that case, equation 6.19 becomes:

$$\begin{aligned} &U \approx Ce^{Qt} \\ &\log(U) \approx \log(C) + (Q/B)t \\ &d(\log(U))/dt \approx Q/B \end{aligned} \tag{6.21}$$

and this case comprises the "fast" set of parameters because the characteristic value of k_1 is relatively large. In this case:

$$U \approx M^*$$

and U is unaffected by changes in k_1, as long as $-k_1 \ll Q$.

The form of the solutions when A is constant is strongly affected by the quantity Y:

$$Y = k_1 + Q = k_1 + k_2 - k_4 - k_3 A \tag{6.22}$$

If $Y \ll 0$, then the quantity U dies away with the rate of k_1/B, and if $Y \gg 0$, then U attenuates with the rate of Q/B. If the magnitude of $|Y|$ is not large, then the solution for U will depend on both rates. If $|Y|$ is small ($\ll 1$), the general solution (equation 6.17) is not reliable because W becomes huge, indeed infinite when $Y = 0$.

When $Y = 0$ (and accordingly, $Q = -k_1$) it is easy to verify that a general solution has the form:

$$\begin{aligned} &M^* = (C + k_1 It)e^{-k_1 t} \\ &U = (C + I + k_1 It)e^{-k_1 t} \end{aligned} \tag{6.23}$$

in which C is an arbitrary constant of integration. In this case, U does not decline at a constant rate, but instead attenuates more slowly. This process loosely resembles the phenomenon called "tailing," which has been observed in some HPP experiments with baro-resistant *L. monocytogenes* (Feeherry et al., 2005).

In summary, for situations when A is constant, the Quasi-chemical model can estimate an appropriate value of the rate, defined as d(log(U))/dt, based on two different sets of parameters, as given above by equation 6.20 ("slow set") and equation 6.21 ("fast set"). In circumstances exhibiting death-only kinetics, such as the inactivation kinetics of *E. coli* by high pressure conditions, the quantity A becomes constant very quickly and the two parameter sets ("slow" and "fast") are almost equally accurate. In contradistinction, A becomes constant relatively late in the process for microbial behaviors that consist of growth-death kinetics, and a unique set of values for the parameters is determined by the behavior during growth and the early stages of death, when A is not constant.

It is important to notice that equation 6.19 and equation 6.23 are accurate approximations for U when A is approximately constant. These circumstances occur near the end of the processes, regardless of whether the microbial behavior consists of growth-death or death-only kinetics. These formulas for U depend not only on the rate constant parameters (the k's), but also on C (the constant of integration), and especially on the constant value of A, which occurs in the definitions of Q, Y, and W. We know that

$$A \leq (k_2/k_3) \tag{6.24}$$

but the value of A can only be found from the numerical solution of the ODE system.

Detection of HPP-Induced Sublethal Injuries

Differential plate counting was carried out in certain experimental conditions by increasing the salinity of the recovery media to discern the extent of injured versus non-injured populations of *E. coli* treated with HPP (Feeherry et al., 2005). As a point of reference, recovering an inoculum of *E. coli* (not treated by HPP) on nutrient agar containing 0–4% added NaCl did not diminish the number of survivors compared to the initial inoculum level. Increasing the NaCl content of the nutrient agar to levels of 4.25% or greater led to decreases in the number of viable cells recovered. Accordingly, a salinity of the recovery media below 4% did not inhibit the growth of healthy cells, but an NaCl content

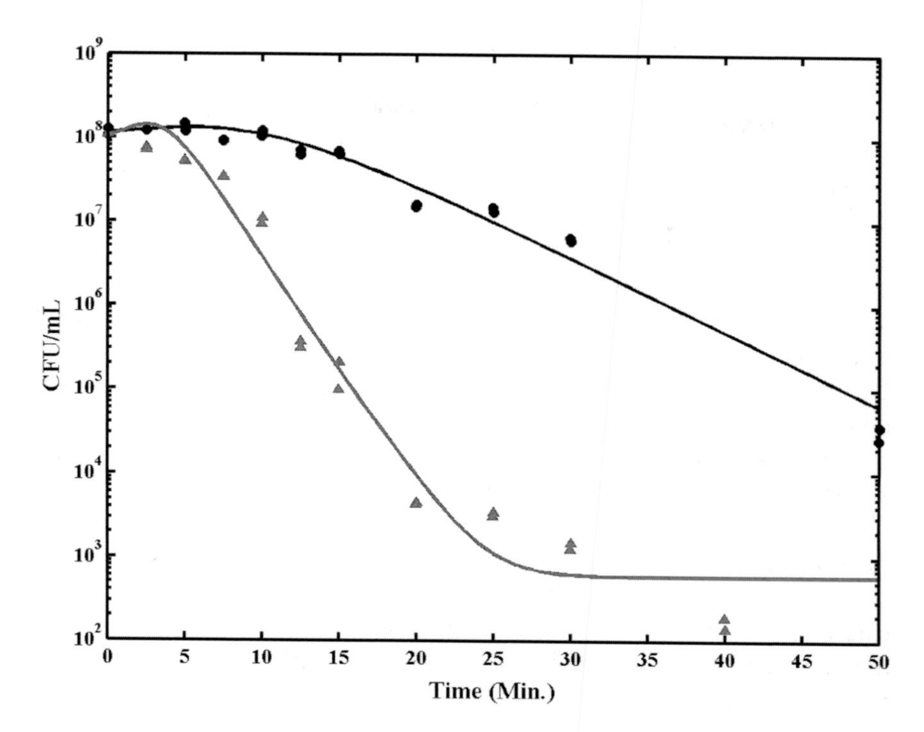

Figure 6.9A. Inactivation kinetics of *E. coli* treated by HPP at 40 kpsi and 40°C and recovered either in regular media (circles, top line—represent recovered injured and uninjured cells) or media containing 4.0% NaCl (triangles, bottom line—represent only uninjured cells) to discern the rapid accumulation of multiple sublethal injuries to the population.

≥ 4.25% imparted an environmental stress to the recovery media that would challenge the survival of weaker, sublethally injured cells.

Whey protein samples inoculated with *E. coli* were exposed to HPP conditions for various times at P = 40 kpsi and T = 40°C (Figure 6.9A). The recovery of HPP-treated *E. coli* on nutrient agar containing 0% added NaCl and on nutrient Agar containing 4% added NaCl revealed sharp differences in the fraction of the population of injured cells versus the fraction of uninjured cells. The enumerated recovery of *E. coli* on nutrient agar containing 0% added NaCl indicated a slow decline in the combined population of uninjured and recoverable, sublethally injured cells (Figure 6.9A, circles—top line). In comparison, the recovery of *E. coli* on nutrient agar containing 4.0% added NaCl enumerated the

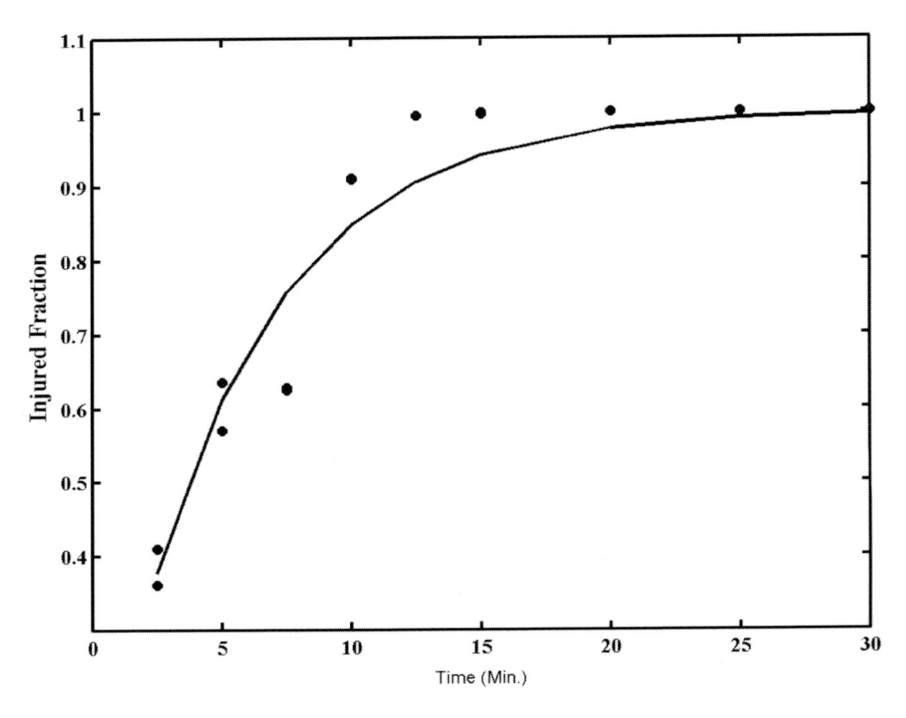

Figure 6.9B. The injured fraction of the population versus time.

population of uninjured cells only, and this population decreased rapidly (Figure 6.9A, triangles—lower line). One surprising aspect observed with this method is that a comparison of the relative rates shows sublethal injury occurs much more rapidly than that of inactivation. After 25 min of HPP, the inoculum exhibited a $<$ 1 log reduction in the total population, but a 5 log reduction in the number of injured and inactivated cells. Figure 6.9B shows that 90% of the population exhibits some form of injury within 12.5 min, which correlates roughly to the lag time (shoulder) observed in the inactivation curve with 0% added NaCl.

Clearly, HPP tends to injure cells before inactivating them, and possibly the accumulation of multiple injuries of increasing severity leads eventually to inactivation of the cells. The inactivation mechanisms of vegetative cells are generally believed to involve critical injuries to cellular organelles such as ribosomes, damage to the cell membrane that leaks vital intracellular constituents, or the inactivation of key enzymes

(Park et al., 2003), although a detailed sequence of steps has not yet been determined.

Concluding Remarks

The present experimental results and their analysis using the Quasi-chemical model and the Weibull model demonstrate the suitability of these alternative models for the interpretation of nonlinear survival curves with particular utility for characterizing the high pressure inactivation of *E. coli* in a surrogate food system. Although these inactivation models derive from different assumptions, they are not mutually exclusive. Rather, they simply address different aspects of what is basically the same phenomenon and operate at different levels of scrutiny. The Quasi-chemical modeling approach seeks to determine the nature and sequence of events occurring at the level of the individual cells, while the "Weibullian approach" tries to capture events at the population level.

Both models depart from convention in their treatment of temperature not as a separate factor, but as an integral part of the coefficients of the inactivation model. How temperature impacts the magnitude of these coefficients was determined experimentally, and not by an a priori assumed relationship. Similarly, the effects of high pressure on the lethality of high temperatures, although not covered in the experimental part of this work, should be accounted for by making the isothermal survival parameters pressure dependent (see equation 6.9 and equation 6.11). The manners in which temperature affects the pressure survival parameters and in which pressure affects the temperature survival parameters also need not be symmetric, because the mechanisms of lethality through which high pressure and high temperature disrupt the cellular machinery are different.

These results show that different mathematical models can feasibly describe the same sets of experimental microbial inactivation data in response to adverse conditions. This applies to any non-isothermal, non-isobaric inactivation model, in which the coefficients of the differential rate equations are functions of the concomitant changes in temperature and pressure (Peleg, 2006). All this derives from the notion that one cannot assume a priori that the temperature and pressure effects are additive or multiplicative, as has been done in other applications in the past. The nature of the high pressure-temperature interaction is determined more plausibly from the organism's response to HPP treatments than from preconceived theory.

In the present study, both the Quasi-chemical model and the Weibull model have a very similar degree of fit of the experimental data, as judged by standard statistical criteria. The real test of their reliability as predictive tools will come when their coefficients, determined from isobaric and isothermal data, are validated in actual foods with experimentally determined survival curves not used in the construction of the models. It would be interesting to determine if the two models discussed in this work could be used to validate survival curves predicted for instances in which both the pressure and temperature vary concomitantly, such as occurs during the come-up times of high pressure treatments. These predicted curves could be tested and compared to subsequent experimental data, to confirm the model's accuracy. If the predictive capabilities of the model are proven correct, then the model could be used to assess the efficacy of postulated treatments, and possibly to suggest ways to improve existing processes. From a theoretical point of view, a compelling area of study would be the determination of whether it is possible to derive all or some of the parameters from each model from those of the other (and vice versa), and directly determine the survival parameters from non-isobaric and non-isothermal experimental data. This will be a real challenge to future investigations in the field and, if proven possible, will result in a comprehensive unified model of microbial inactivation by lethal temperature and pressure combinations.

Future work will build on these results to include other types of nonlinearities such as "tailing," as exhibited by baro-resistant strains of vegetative pathogens (Tay et al., 2003) of *L. monocytogenes* and tough-to-kill bacterial endospores to achieve commercial sterility of foods with HPP. We have observed "tailing" in some sets of conditions with *L. monocytogenes* as the target pathogen (Feeherry et al., 2005). In these cases, the original four-parameter Quasi-chemical model was not sufficient to account for the observed phenomena. We have suggested plausible modifications to the four-parameter model to accommodate these observations, particularly the addition of a fifth parameter (k_5) that counteracts the effects of k_4. The introduction of this fifth parameter represents a reasonable starting point to account for tailing with the Quasi-chemical model and bodes well for the future use of this model in describing the nonlinear inactivation kinetics of tough-to-kill bacterial spores. The Quasi-chemical model and the Weibull model need further refinement in order to predict the safety of foods in industrial HPP applications. Presently, these models provide a new conceptual framework of how to deal with nonlinear inactivation kinetics in a manner that is consistent with the actual events that take place at the levels of the microbial cell and the population.

References

ABC News. 2002, 22 May. World News Tonight. ABC News, Inc.

Ahn, J., V.M. Balasubramaniam, and A.E. Yousef. 2007. Inactivation kinetics of selected aerobic and anaerobic bacterial spores by pressure-assisted thermal processing. *International Journal of Food Microbiology* 113:321–329.

Baranyi, J., and T.A. Roberts. 1997. Mathematics of predictive food microbiology. *International Journal of Food Microbiology* 24:199–218.

———. 2000. "Principles and applications of predictive modeling of the effects of preservative factors on microorganisms." In: *The Microbiological Safety and Quality of Food, Vol 1,* ed. B.M. Lund, T.C. Baird-Parker, and G.W. Gould, pp. 342–358. Gaithersburg, MD: Aspen Publishers.

Bassler, B.L. 1999. How bacteria talk to each other: Regulation of gene expression by quorum sensing. *Current Opinions in Microbiology* 2(6):582–587.

BBC News. 2002. US military unveils "super sandwich." Available from http://news.bbc.co.uk/hi/english/world/americas/newsid_1923000/1923054.stm. Accessed 11 April 2002.

Bray, H.G., and K. White. 1966. *Kinetics and Thermodynamics in Biochemistry*, pp. 364–404. New York: Academic Press.

Buchanan, R.L. 1992. "Predictive microbiology: Mathematical modeling of growth in foods." In: *American Chemical Society Symposium Series. Food Safety Assessment,* ed. J.W. Finley, S.F. Robinson, and D.J. Armstrong, 484:250–260.

Buchanan, R.L., M.H. Golden, and R.C. Whiting. 1993. Differentiation of the effects of pH and lactic or acetic acid concentration on the kinetics of *Listeria monocytogenes* inactivation. *Journal of Food Protection* 56(6):474–478.

Chen, H. 2007. Use of linear, Weibull, and log-logistic functions to model pressure inactivation of seven foodborne pathogens in milk. *Food Microbiology* 24:197–204.

Chen, H., and D.G. Hoover. 2004. Use of Weibull model to describe and predict pressure inactivation of *Listeria monocytogenes* Scott A in whole milk. *Innovative Food Science and Emerging Technologies* 5:269–276.

CNN.com. 2002. Soldiers snack on 3-year sandwich. Available from http://www.cnn.com/2002/TECH/science/04/11/us.food/index.html. Accessed 11 April 2002.

Cook, G. 2002, 13 April. Army develops sandwich to stand test of time. *Boston Globe.* Available from http://www.boston.com/globe/.

Del Nobile, M.A., D. D'Amato, C. Altieri, M.R. Corbo, and M. Sinigaglia. 2003. Modeling the yeast growth-cycle in a model wine system. *Journal of Food Science* 68(6):2080–2085.

Doona, C.J., F.E. Feeherry, and M.-Y. Baik. 2006. Water dynamics and retrogradation of ultrahigh-pressurized wheat starch. *Journal of Agricultural and Food Chemistry* 54:6719–6724.

Doona, C.J., F.E. Feeherry, and E.W. Ross. 2005. "A Quasi-chemical model for growth and death of microorganisms in foods by non-thermal and high pressure processing." In: *4th International Conference Proceedings of Predictive Modeling in Foods. International Journal of Food Microbiology,* ed. J.F.M. Van Impe, A.H. Geeraerd, I. Legeurinel, and P. Mafart, 100:21–32.

Dunny, G.M., and S.C. Winans (eds.). 1999. *Cell–Cell Signaling in Bacteria.* Washington, DC: American Society for Microbiology Press.

England, R.R., G. Hobbs, N.J. Bainton, and D.M. Roberts. 1999. *Microbial Signaling and Communication, 57th Symposium of the Society for General Microbiology.* Cambridge, UK: Cambridge University Press.

Epstein, I.R., and J.A. Pojman. 1998. *An Introduction to Nonlinear Chemical Dynamics.* New York: Oxford University Press.

Fábián, I., and R. van Eldik. 1993. Complex-formation kinetics of iron(III) with chlorite ion in aqueous solution. Mechanistic information from pressure effects. *Inorganic Chemistry* 32(15):3339–3342.

Fabricant, F. 2002, 1 May. FOOD STUFF; Hardtack gets a battlefield promotion. *New York Times.* Available from http://www.nytimes.com.

Feeherry, F.E., C.J. Doona, and E.W. Ross. 2005. The Quasi-chemical kinetics model for the inactivation of microbial pathogens using high pressure processing. *Acta Horticulturae* 674:245–251.

Feeherry, F.E., E.W. Ross, and I.A. Taub. 2001. Modeling the growth and death of bacteria in intermediate moisture foods. *Acta Horticulturae* 566:123–128.

Feeherry, F.E., I.A. Taub, and C.J. Doona. 2003. Effect of water activity on *Staphyloccocus aureus* growth in intermediate moisture bread. *Journal of Food Science* 68:982–987.

Fussman, G.F., S.P. Ellner, K.W. Shertzer, and N.G. Hairston. 2000. Crossing the Hopf bifurcation in a live predator-prey system. *Science* 290:1358–1360.

Goldbeter, A. 1991. Biochemical oscillations and cellular rhythms. London: Cambridge University Press.

———. 1996. Biochemical oscillations and cellular rhythms: The molecular bases of periodic and chaotic behaviour. Cambridge, UK: Cambridge University Press.

Graham-Rowe, D. 2002. US military creates indestructible sandwich. Available from http://www.newscientist.com/news/print.jsp?id=ns99992151. Accessed 10 April 2002.

Guan, D., H. Chen, E.Y. Ting, and D.G. Hoover. 2006. Inactivation of *Staphylococcus aureus* and *Escherichia coli* O157:H7 under isothermal-endpoint pressure conditions. *Journal of Food Engineering* 77:620–627.

Heldman, D.R., and R.L. Newsome. 2003. Kinetic models for microbial survival during processing. *Food Technology* 57(8):40–46 and 100.

Hinshelwood, C.N. 1953. Autosynthesis. *Journal of the Chemical Society* 1947–1956.

Jones, J.E., and S.J. Walker. 1993. Advances in modeling microbial growth. *Journal of Industrial Microbiology* 12:200–205.

Jones, J.E., S.J. Walker, J.P. Sutherland, M.W. Peck, and C.L. Little. 1994. Mathematical modeling of the growth, survival, and death of *Yersinia enterocolitica*. *International Journal of Food Microbiology* 23:433–447.

Kamau, D.N., S. Doores, and K.M. Pruitt. 1990. Enhanced thermal destruction of *Listeria monocytogenes* and *Staphylococcus aureus* by the lactoperoxidase system. *Applied and Environmental Microbiology* 56(9):2711–2716.

Kleerebezem, M., L.E.N. Quadri, O.P. Kuipers, and W.M. de Vos. 1997. Quorum sensing by peptide pheromones and two-component signal transduction systems in gram-positive bacteria. *Molecular Microbiology* 24:895–904.

Legan, D., M. Vandeven, C. Stewart, and M. Cole. 2002. "Modeling the growth, survival, and death of bacterial pathogens in foods." In: *Foodborne Pathogens: Hazards, Risk Analysis, and Control,* ed. C. de W. Blackburn and P.J. McClure, pp. 53–95. Cambridge, UK: Woodhead Publishing.

Margosch, D., M.A. Ehrmann, R. Buckow, V. Heinz, R.F. Vogel, and M.G. Gänzle. 2006. High-pressure-mediated survival of *Clostridium botulinum* and *Bacillus amyloliquifaciens* endospores at high temperature. *Applied and Environmental Microbiology* 72(5):3476–3481.

Margosch, D., M.A. Ehrmann, M.G. Gänzle,, and R.F. Vogel. 2004a. Comparison of pressure and heat resistance of *Clostridium botulinum* and other endospores in mashed carrots. *Journal of Food Protection* 67(11):2530–2537.

Margosch, D., M.G. Gänzle, M.A. Ehrmann, and R.F. Vogel. 2004b. Pressure inactivation of *Bacillus* endospores. *Applied and Environmental Microbiology* 70(12):7321–7328.

McMeekin, T.A., J. Brown, K. Krist, D. Miles, K. Neumeyer, D.S. Nichols, J. Olley, K. Presser, D.A. Ratkowsky, T. Ross, M. Salter, and S. Soontranon. 1997. Quantitative microbiology: A basis for food safety. *Emerging and Infectious Diseases* 3(4):541–549.

McMeekin, T.A., J.N. Olley, T. Ross, and D.A. Ratkowsky. 1993. *Predictive Microbiology: Theory and Application,* ed. A.N. Sharpe. Somerset, England: Research Studies Press LTD.

Membré, J.M., J. Thurette, and M. Catteau. 1997. Modeling the growth, survival, and death of *Listeria monocytogenes*. *Journal of Applied Microbiology* 82:345–350.

Novick, R.P. 1999. "Regulation of pathogenicity in *Staphylococcus aureus* by a peptide-based density-sensing system." In: *Cell-Cell Signaling in Bacteria,* ed. G.M. Dunny and S.C. Winans, pp. 129–146. Washington, DC: American Society for Microbiology Press.

Park, S.-J., H.-W. Park, and J. Park. 2003. Inactivation kinetics of food poisoning microorganisms by carbon dioxide and high hydrostatic pressure. *Journal of Food Science* 68(3):976–981.

Peleg, M. 1996. A model of microbial growth and decay in a closed habitat based on a combined Fermi's and the logistic equation. *Journal of Science and Food Agriculture* 71:225–230.

———. 2003. Microbial survival curves: Interpretation, mathematical modeling and utilization. *Comments on Theoretical Biology* 8:357–387.

———. 2006. *Advanced Quantitative Microbiology for Foods and Biosystems: Modeling and Predicting Growth and Inactivation*. Boca Raton, FL: CRC Press.

Peleg, M., and M.B. Cole. 1998. Re-interpretation of microbial survival curves. *Critical Reviews in Food Science and Nutrition* 38:353–380.

Peleg, M., and C.M. Penchina. 2000. Modeling microbial survival during exposure to a lethal agent with varying intensity. *Critical Reviews in Food Science and Nutrition* 40:159–172.

Ross, E.W., I.A. Taub, C.J. Doona, F.E. Feeherry, and K. Kustin. 2005. The mathematical properties of the Quasi-chemical model for microorganism growth/death kinetics in foods. *International Journal of Food Microbiology* 99:157–171.

Skinner, G.E., J.W. Larkin, and E.J. Rhodehamel. 1994. Mathematical modeling of microbial growth: A review. *Journal of Food Safety* 14:175–217.

Smith, J.L., P.M. Fratamico, and J.S. Novak. 2004. Quorum sensing: A primer for food microbiologists—Review. *Journal of Food Protection* 67(5):1053–1070.

Taub, I.A, F.E. Feeherry, E.W. Ross, K. Kustin, and C.J. Doona. 2003. A Quasi-chemical kinetics model for growth and death of *Staphylococcus aureus* in intermediate moisture bread. *Journal of Food Science* 68:2530–2537.

Tay, A., T.H. Shellhammer, A.E. Yousef, and G.W. Chism. 2003. Pressure death and tailing behavior of *Listeria monocytogenes* strains having different barotolerances. *Journal of Food Protection* 66(11):2057–2061.

van Boekel, M.A.J.S. 2002. On the use of the Weibull model to describe thermal inactivation of microbial vegetative cells. *International Journal of Food Microbiology* 74:139–159.

Whiting, R.C. 1993. Modeling bacterial survival in unfavorable environments. *Journal of Industrial Microbiology* 12:240–246.

———. 1995. Microbial modeling in foods. *Critical Reviews in Food Science and Nutrition* 35(6):467–494.

Whiting, R.C., and R.L. Buchanan. 1994. Scientific status summary—Microbial modeling. *Food Technology* 48(6):113–120.

Whiting, R.C., and M.L. Cygnarowicz-Provost. 1992. A quantitative model for bacterial growth and decline. *Food Microbiology* 9:269–277.

Whiting, R.C., S. Sackitey, S. Calderone, K. Morely, and J.G. Phillips. 1996. Model for the survival of *Staphylococcus aureus* in nongrowth environments. *International Journal of Food Microbiology* 31:231–243.

Yahoo! News. 2002. New secret weapon—The indestructible sandwich. Available from http://story.news.yahoo.com/news?tmpl+story&u=/nm/20020411/od_nm/sandwiches_dc_1. Accessed 11 April 2002.

Chapter 7

Sensitization of Microorganisms to High Pressure Processing by Phenolic Compounds

Yoon-Kyung Chung, Aaron S. Malone, and Ahmed E. Yousef

Introduction

The production of spore-free food through commercial sterilization is normally associated with thermal methods that result in the over-processing of foodstuffs. Foods processed in this manner suffer losses of nutrients and of overall quality that render it inferior to their fresh counterpart. Achieving commercial sterility without over-processing is a noble goal, but the means to accomplishing this goal are limited. Currently, alternative effective sterilization procedures are lacking. High pressure processing (HPP) is a promising alternative treatment, but the process per se has limited or no efficacy against bacterial spores. Nakayama et al. (1996) reported that ultra-high pressure sterilization of low-acid foods was possible only when high pressure was combined with high temperatures. Additionally, survivors among these treated spores did not germinate simultaneously. This suggests that pressure sterilization, even at high temperatures, is inherently unreliable. Food processors often combine lethal treatments to assure food safety. Process combinations that act synergistically against bacterial spores are most desirable, and improvements in spore inactivation by HPP have been achieved when the pressure treatments were used in conjunction with bacteriocins and chemical preservatives (Shearer et al., 2000; López-Pedemonte et al., 2003; Cléry-Barraud et al., 2004).

Many pathogenic vegetative microorganisms that are of concern to food processors are also sensitive to HPP (Lado and Yousef, 2002).

Therefore, high pressure is currently used in lieu of thermal pasteurization to process fruit juices, purees, guacamole, desserts, sauces, oysters, rice dishes, and packaged cured ham (Grant et al., 2000). Applying moderate conditions of HPP produces food retaining its desirable qualities, but such treatment alone may not sufficiently decrease foodborne disease hazards to acceptable levels. Some food additives increase the lethality of pressure against the microbial load in food. Phenolic compounds, at levels innocuous to microorganisms, may in fact sensitize them to the high pressure treatment. In this chapter, "sensitization" is defined as any sublethal or mildly lethal treatment that increases the efficacy of a subsequent bactericidal process. Therefore, the sensitizer, at the level applied, should have no, or only limited, bactericidal action. This definition implies that a sensitizer may have a bacteriostatic (inhibitory) effect against the targeted microorganism. A phenolics-pressure combination is potentially useful in controlling pathogens, especially in fat-rich foods such as salad dressings and sausages. This synergy not only improves the safety of the product, it also allows food processors to use milder and economical pressure treatments. Although studies depicting this synergy against bacterial spores are lacking, combining HPP and phenolics is a promising strategy to improve the safety of minimally or nonthermally processed foods.

Phenolic Compounds

Chemistry of Phenolics

Derivatives of phenol, called phenolics, contain a phenol moiety with one or more substitutions. Anti-microbial phenolics, that are relevant to food, can be classified as (i) currently approved preservatives (i.e., parabens), (ii) additives approved for uses non-related to their anti-microbial properties (e.g., antioxidants), and (iii) natural food ingredients (e.g., isoeugenol, carvacrol, catechins). Parabens (alkyl esters of *p*-hydroxybenzoic acid) are allowed in many countries as anti-microbial food additives. The phenolic antioxidants butylated hydroxytoluene (BHT), butylated hydroxyanisole (BHA), propyl gallate (PG), and *tert*-butylhydroquinone (TBHQ) are additives used to prevent oxidative rancidity of lipids in food (Shahidi et al., 1992). Naturally occurring phenolic compounds are widespread in plants and food systems. These include phenolic derivatives (e.g., *p*-cresol, vanillic acid, gallic acid, hydroquinone), hydroxycinnamic acid derivatives (e.g., *p*-coumaric, caffeic, ferulic and sinapic acids), flavonoids (e.g., catechins, proanthocyanins, anthocyanidins and flavons, flavonols and their glycosides), and tannins (e.g., plant polymeric phenolics with the ability to precipitate protein from aqueous solutions). Polyphenolic compounds (flavonoids, catechols, and derivatives of gallic acid) are widely distributed in edible plants and considered to be dietary antioxidants (López-Malo et al., 2005). The structures of some of these phenolics are shown in Table 7.1. These selected phenolics are emphasized in this chapter and some have exhibited anti-microbial activity.

Anti-microbial Activity of Phenolics

Parabens

Parabens have been used in foods for more than 50 years because of their low toxicity and effectiveness as anti-microbial agents. Parabens, in general, are more effective against fungi than bacteria, with a tendency to be more effective against Gram-positive than Gram-negative bacteria (Soni et al., 2002). The anti-microbial activity of parabens increases as the length of alkyl chain increases (Davidson, 2005). Draughon et al. (1982) demonstrated that 1,000 ppm of four parabens (i.e., methyl-, ethyl-, propyl-, and butyl-parabens) prevented the growth of *Clostridium botulinum* spores inoculated in fluid thioglycollate broth for 24 hr at 35°C. These authors also showed that ethyl-paraben at 1,000 ppm inhibited the growth and toxin formation by *C. botulinum*, which were inoculated at 100 spores/g of meat in canned pork, during 4 weeks of incubation. However, inhibition of *C. botulinum* by parabens has been reported to be much less effective in foods than in laboratory media (Sofos and Busta, 1980). Propylparaben (at 250 ppm) did not prevent the growth of *Listeria monocytogenes* inoculated on vacuum-packaged sliced ham or sausage samples (Blom et al., 1997). Numerous studies tested the anti-microbial efficacy of parabens in laboratory media and foods, but only a limited number only of these studies are reported in this chapter.

Phenolic Antioxidants

Phenolics such as BHA, BHT, and TBHQ are used to prevent the autooxidation of fat in food. These phenolic antioxidants occasionally function

Table 7.1. Structure of selected phenolics relevant to food preservation

Classification	Structure
"Generally recognized as safe" food antimicrobials	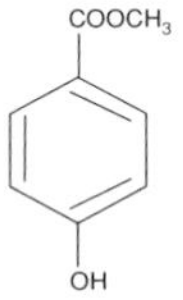 Methyl paraben; 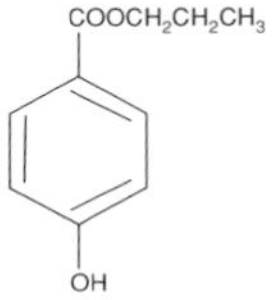 Propyl paraben
Approved food antioxidants	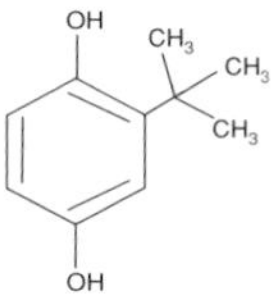 TBHQ; 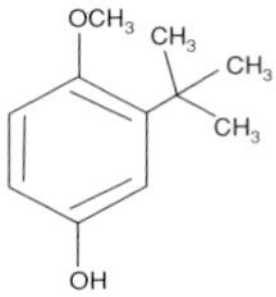 BHA; 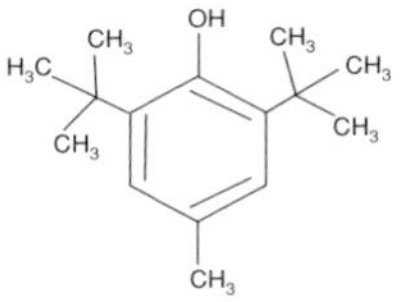 BHT; 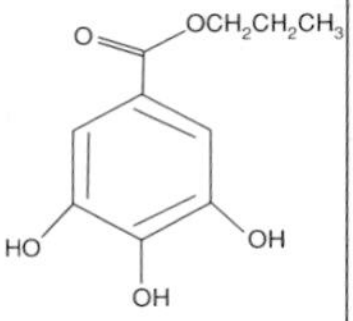 PG
Naturally occurring plant compounds: phenolic derivatives	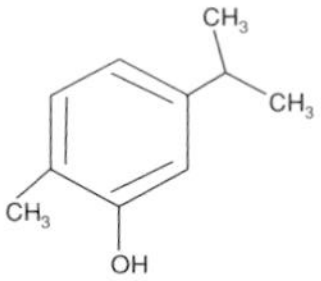 Carvacrol; 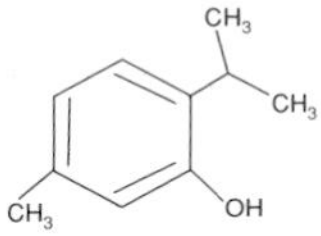 Thymol; 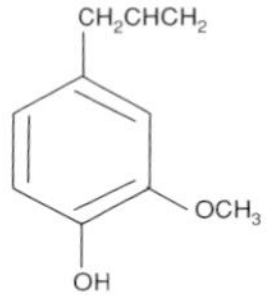 Eugenol; 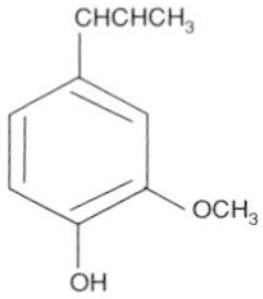 Isoeugenol; 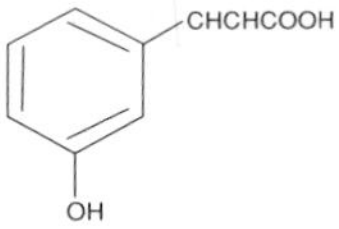 Caffeic acid; 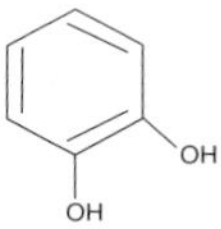 Catechol; 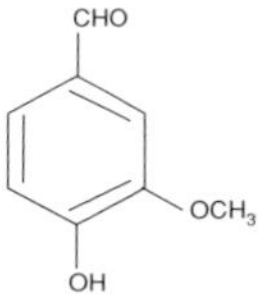 Vanillin; 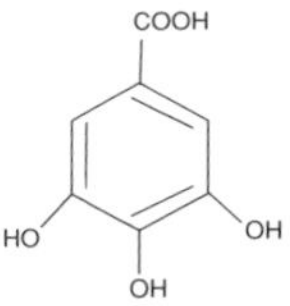 Gallic acid
Naturally occurring plant compounds: flavonoids and polyphenolics	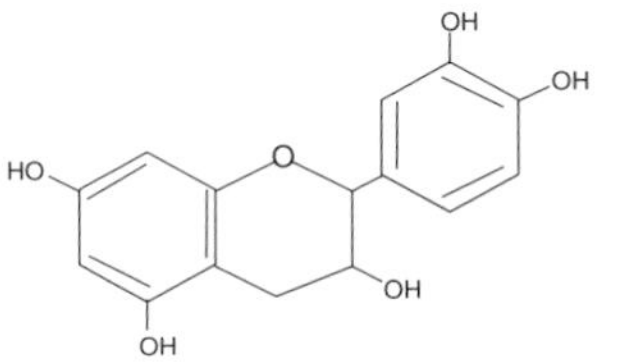 Catechin; 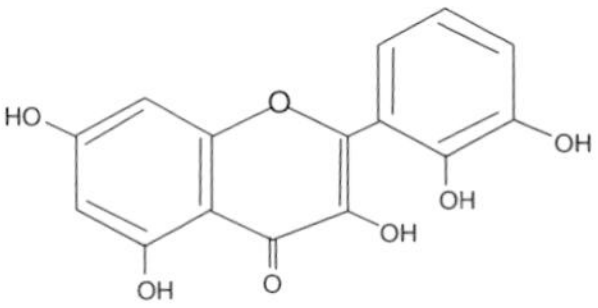 Quercetin; 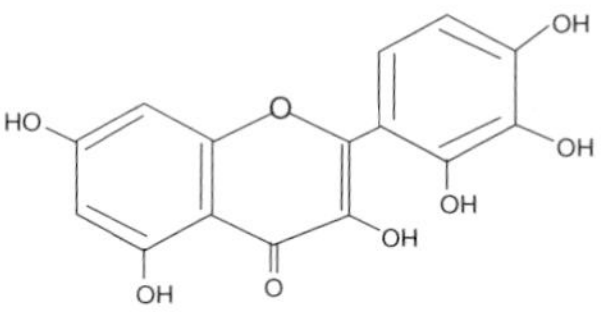 Myricetin; 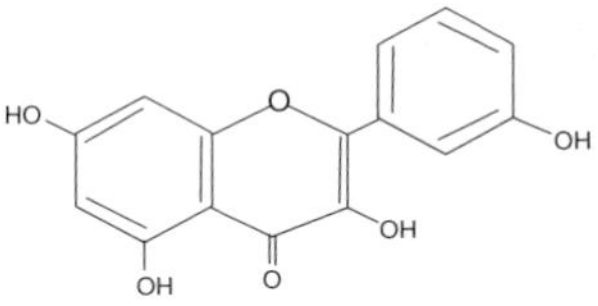 Kaempherol

as anti-microbial agents. Studies indicated that BHA (at 50–400 ppm) inhibited the growth of various microorganisms including *Staphylococcus aureus*, *Escherichia coli*, *Vibrio parahaemolyticus*, *Clostridium perfringens*, and some species of *Salmonella* and *Pseudomonas* (Chang and Branen, 1975; Robach et al., 1977; Davidson et al., 1979; Klindworth et al., 1979; Davidson and Branen, 1980a, 1980b). Yousef et al. (1991) also indicated that BHA (100–300 ppm), BHT (300–700 ppm), and TBHQ (10–30 ppm) inhibited the growth of *L. monocytogenes* in tryptose broth, showing mostly a bacteriostatic rather than bacteriocidal action. Few investigators tested the anti-microbial activity of these phenolic antioxidants against bacterial spores. Reddy et al. (1982) reported that BHT, BHA, and TBHQ (at 200–400 ppm) inhibited the growth and toxin production by *C. botulinum* spores (2.5×10^2 spores/ml) in thiotone-yeast extract-glucose broth, for 7 days.

Naturally Occurring Phenolic Compounds

Plants contain anti-microbial compounds in the essential oil fraction of the leaves (e.g., rosemary, sage), flowers and flower buds (e.g., clove), and other organs. The phenolic components of these essential oils are chiefly responsible for the anti-microbial properties. Deans and Ritchie (1987) examined fifty plant essential oils for their antibacterial properties against twenty-five genera of bacteria. Among these, some of the most effective oils were thyme, cinnamon, bay, clove, and marjoram. Recently, the anti-microbial activity of essential oils has been extensively reviewed (Burt, 2004). Some of the studies testing the efficacy of natural phenolics against bacterial spores are briefly included here. Ismaiel and Person (1990b) reported that 100 ppm cinnamon, thyme, or origanum oils, or 150 ppm clove or pimenta oils prevented the germination of *C. botulinum* spores. Origanum oil, at 200 ppm, was effective in delaying the growth and toxin production of *C. botulinum* spores (2.5×10^2 spores/ml) in thiotone-yeast extract-glucose broth (Reddy et al., 1982). However, the antibotulinal effect of origanum oil was dramatically reduced in the pork sample, and a significant effect was only shown when the additive was used at 400 ppm, in combination with 50–100 ppm sodium nitrite (Ismaiel and Pierson, 1990a).

Chaibi et al. (1997) tested nine plant essential oils (eucalyptus, camomile, cedar, savage carrots, artemisia, rosemary, orange, grapefruit, and vervain) against *C. botulinum* and *Bacillus cereus* spores. The variability in anti-microbial activities of essential oils has been ascribed to the differences in their active fraction. Essential oils have the most inhibitory activity when they contain phenolic compounds (Kurita et al., 1981; Gueldner et al., 1985). According to Chaibi et al. (1997), rosemary oil, which contains phenolics as the main active components, was sporicidal against *B. cereus* spores at a lower concentration ($\geq$ 170 ppm) than other essential oils with non-phenolic active components; the latter showed sporicidal effects at $\geq$ 300 ppm. This study also demonstrated that these essential oils inhibit specifically one or more stages of the spore transformation cycle depending on the nature and the concentration of essential oils and bacterial spore strains. Al-Khayat and Blank (1985) indicated that eugenol, a major component in cloves, was sporostatic at 500–600 ppm to *B. subtilis* spores. Other related phenolics, including isoeugenol, gingerol, and zingerone, also were sporostatic at $\geq$ 500, $\geq$ 9000, and $\geq$ 8,000 ppm, respectively. In addition, the sporostatic activity of phenols and phenolic derivatives was noticed in earlier studies (Sierra, 1970; Jurd et al., 1971).

Mode of Anti-microbial Action of Phenolics

Although the exact anti-microbial mechanism of phenolic compounds is not well established, it is generally believed that these compounds disrupt the cellular cytoplasmic membrane and impair its related functions (i.e., proton motive force, electron flow, and active transport), or coagulate cell contents (Denyer and Hugo, 1991; Sikkema et al., 1995; Davidson, 1997). Considering the similarities in structures of phenolics, it seems plausible that their mechanism of action would therefore be related to that of phenol. Fogg and Lodge (1945) first proposed that the anti-microbial activity of phenol was due to the inactivation of cellular enzymes and that the rate limiting step was the penetration of the compound into the cell. Phenolics increased membrane permeability, due to a weakening or destruction of the permeability barrier of the cell membrane, in *E. coli* (Judis, 1963) and *Pseudomonas aeruginosa* (Bernheim, 1972). It was proposed that phenol reacted primarily with the membrane phospholipids of *P. aeruginosa*, subsequently causing an increase in the permeability of the cell membrane (Bernheim, 1972). Furr and Russell (1972) observed a leakage of RNA from *Serratia marcescens* in the presence of parabens, and reported that the amount of leaked RNA was proportional to the alkyl chain length of the paraben. Juneja and

Davidson (1993) reported that *L. monocytogenes*, grown in the presence of exogenously added C14:0 or C18:0 fatty acids, was resistant to TBHQ and parabens, while the cells grown with C18:1 fatty acid were sensitive to these compounds. These authors indicated that a possible correlation existed between the lipid composition of the cell membrane and the susceptibility of *L. monocytogenes* to these anti-microbial agents. Rico-Munoz et al. (1987) examined the effect of BHA, BHT, TBHQ, PG, *p*-coumaric acid, ferulic acid, caffeic acid, and methyl- and propyl-parabens on the membrane-bound ATPase of *S. aureus*. Propyl gallate, TBHQ, and *p*-coumaric, ferulic, and caffeic acids inhibited the ATPase activity, while BHA was found to stimulate the activity of the enzyme. Neither BHT nor parabens affected the enzyme. Therefore, the authors concluded that the phenolic compounds probably did not have the same cell target or mechanism.

Essential oils that possess the strongest antibacterial properties against foodborne pathogens generally contain a relatively high concentration of phenolic compounds such as carvacrol, eugenol, and thymol (Dorman and Deans, 2000). The mode of action of carvacrol, one of the major components of oregano and thyme oils, has been studied most among the essential oil compounds (Ultee et al., 1998, 1999, 2000, 2002; Lambert et al., 2001). An important characteristic for antibacterial properties of essential oil components is their hydrophobicity, which enables them to partition in the lipids of cell membrane. Membrane-accumulated carvacrol occupies more than the normal spaces between fatty acid chains of the phospholipids bi-layer, and results in conformational change and fluidization of the membrane (Ultee et al., 2000, 2002). Similarly, Ultee et al. (1999) observed that carvacrol interacted with the phospholipid bi-layer in *B. cereus* cell membrane, dissipated the membrane potential, and consequently increased the permeability of cytoplasmic membrane for protons and potassium ions. Ultee et al. (2000) concluded that carvacrol formed channels through the membrane, and it allowed ions to leave the cytoplasm. Lambert et al. (2001) also observed that oregano essential oil, which contained carvacrol as a major component, caused the leakage of phosphate ions from *S. aureus* and *P. aeruginosa*.

The chemical structure of the essential oil components also affects their antibacterial activity and mode of action. The importance of the hydroxyl group in phenolic compounds such as carvacrol for their antimicrobial activity has been confirmed (Dorman and Deans, 2000; Ultee et al., 2002). Dorman and Deans (2000) observed differences in antimicrobial efficacy between carvacrol and thymol against Gram-negative and Gram-positive bacteria, suggesting that the relative position of the hydroxyl group in the phenolic compounds also contributed to the antibacterial efficacy of these compounds.

Tsuchiya (1999) tested eight catechins (catechin, gallocatechin, catechin gallate, gallocatechin gallate, epicatechin, epigallocatechin, epicatechin gallate, epigallocatechin gallate) for their effects on membrane fluidity using liposome. All of these catechins, ranging from 1 to 1000 μ mol/L, significantly decreased membrane fluidity in both hydrophilic and hydrophobic regions of lipid bi-layers. Sakanaka et al. (2000) tested a mixture of green tea polyphenols (catechin, gallocatechin, gallocatechin gallate, epicatechin, epicatechin gallate, epigallocatechin gallate) for the effect against bacterial spores. The addition of 500 ppm tea polyphenols significantly decreased the heat resistance of *B. stearothermophilus* and *C. thermoaceticum* spores, with the latter more sensitized to heat than the former. Heating *C. thermoaceticum* spores at 120°C for 50 min decreased their population by 6 logs in their presence, and 1 log only in their absence of 500 ppm polyphenols, respectively (Sakanaka et al., 2000). Polyphenolic compounds are commonly known to interact with proteins (Arts et al., 2002; Papadopoulou and Frazier, 2003; Chen and Hagerman, 2004). The effect of tea polyphenols on the reduction of heat resistance of spores may be caused by the interaction between tea phenols and spore coat proteins (Sakanaka et al., 2000).

Combination of High Pressure Processing and Phenolics

High pressure processing often results in microbial inactivation patterns that leave a fraction of the remaining population viable, even after prolonged processing. HPP may be combined with other preservation methods to increase its efficacy and commercial feasibility. Currently used and potential food additives, such as bacteriocins (Yuste et al., 2002), potassium sorbate (Mackey et al., 1995), and carvacrol (Karatzas et al., 2001), have been tested in combination with HPP. Combining phenolic food-grade ingredients with HPP is likely useful in eradicating pressure-resistant microorganisms in food. Included in this chapter are our recent efforts to identify phenolic compounds to enhance the efficacy of high pressure against pathogenic or spoilage microorganisms including *L. monocytogenes*, *E. coli* O157:H7, and *Lactobacillus plantarum* (Chung et al., 2005; Vurma et al., 2006; Malone et al., unpublished data, 2006). The reader should be cautioned that these studies targeted mainly non-sporulating bacteria, and the sensitization of bacterial spores to high pressure by phenolics has not been investigated in a systematic fashion.

Comparing Phenolics for Synergy with High Pressure

Three strains of *L. monocytogenes* (Scott-A, OSY-8578, and OSY-328) that vary considerably in pressure resistance and susceptibility to additives (particularly nisin) were tested (Vurma et al., 2006). The strains were treated, individually, with 100 ppm phenolic compounds, followed by pressurization at 400 MPa for 5 min. Although the phenolic compounds alone were not lethal to *L. monocytogenes*, these compounds significantly ($p < 0.0001$) enhanced the pressure lethality against the pathogen (Table 7.2). Sensitization of *L. monocytogenes* to pressure by additives followed this descending order: TBHQ, BHT, carvacrol, BHA, PG, and phenol. Isoeugenol, hydroquinone, trihydroxybutyrophenone, catechin, thymol, and rosemary extract did not sensitize *L. monocytogenes* to pressure. The sensitizing action of TBHQ was significantly greater ($p = 0.02$) than that of the other additives tested. According to an earlier study, BHA increased the lethality of pressure against *L. monocytogenes* by 4 log, but BHT showed no sensitizing effect (Mackey et al., 1995). Karatzas et al. (2001) also reported a synergistic action between carvacrol and HPP against *L. monocytogenes* Scott-A in buffer and in low-fat milk.

Phenolic-Pressure Treatment Sequence

The sensitizing action of phenolics was evaluated when these compounds were added before or after the pressure treatment (Vurma et al., 2006). The addition of TBHQ increased the lethality of HPP against *L. monocytogenes* regardless of the application sequence (Table 7.3). However, the lethality of the combination treatment was greater when TBHQ was added before, rather than after, the pressurization at both 400 and 500 MPa. In contrast, Karatzas et al. (2001) reported that the

Table 7.2. Inactivation of *Listeria monocytogenes*[a] by treatments with high pressure (400 MPa, 5 min, 18–20°C) and selected phenolic compounds in phosphate buffer and dimethyl/sulfoxide solutions (adapted from Vurma et al., 2006)

Treatment Combinations[b]	Log CFU/mL Decreased[c]	Significance of Selected Contrasts[d]	
		Contrast	Probability
Additives only	0.08	Additives vs. non-treated (control)	0.75
HPP only	2.4	HPP vs. additives	<0.0001
		HPP vs. additives-HPP	<0.0001
BHA-HPP	3.4	HPP vs. BHA-HPP	0.002
BHT-HPP	3.7	HPP vs. BHT-HPP	<0.0001
Carv-HPP	3.6	HPP vs. Carv-HPP	0.0001
PG-HPP	3.3	HPP vs. PG-HPP	0.004
Phe-HPP	3.2	HPP vs. Phe-HPP	0.007
TBHQ-HPP	4.0	HPP vs. TBHQ-HPP	<0.0001
		TBHQ-HPP vs. other additives-HPP	0.020

[a] Three *L. monocytogenes* strains (Scott-A, OSY-8578, and OSY-328) were tested; initial population was 10^9 CFU/mL.
[b] HPP, high pressure processing; BHA, butylated hydroxyanisole; BHT, butylated hydroxytoluene; Carv, carvacrol; PG, propyl gallate; Phe, phenol; TBHQ, *tert*-butylhydroquinone. Additional phenolics (isoeugenol, hydroquinone, trihydroxybutyrophenone, catechin, thymol, and rosemary extract) did not sensitize *L. monocytogenes* to pressure (data not shown).
[c] Average for four independent trials.
[d] Means were contrasted using the general linear model of SAS statistical program.

effect of adding carvacrol after HPP was not significantly different from that of the simultaneous application of HPP and carvacrol.

Strain Variability to Phenolic-Pressure Treatments

Microorganisms vary considerably in resistance to pressure. This variability is commonly observed not only among genera and species, but also among strains of the same species (Alpas et al., 1999; Tay et al., 2003; Malone et al., 2006). The variation in strain sensitivity to a certain process is an important consideration when designing processing

Table 7.3. Populations of *Listeria monocytogenes*[a] that survived processing combinations when the order of applying high pressure (HPP) and treating with *tert*-butylhydroquinone (TBHQ) was changed (adapted from Vurma et al., 2006)

Pressure (MPa)	Sequence of Treatment	Log CFU/ml[b]	Significance of selected contrasts[c]	
			Contrast	Probability
None	None	8.9		
400	HPP only	4.5		
	HPP then TBHQ	3.2	HPP-TBHQ vs. HPP	<0.0001
	TBHQ then HPP	2.1	TBHQ-HPP vs. HPP-TBHQ	0.0013
500	HPP only	3.7		
	HPP then TBHQ	2.6	HPP-TBHQ vs. HPP	0.0004
	TBHQ then HPP	1.3	TBHQ-HPP vs. HPP-TBHQ	0.0002
400 and 500			TBHQ-HPP vs. HPP-TBHQ	< 0.0001

[a] Three *L. monocytogenes* strains (Scott-A, OSY-8578, and OSY-328) were tested.
[b] Average for three independent trials.
[c] Means were contrasted using the general linear model of SAS statistical program.

technologies. These technologies should be effective against the most resistant strain of the microorganism of concern.

Variation in sensitivity to pressure-phenolic treatments was tested using *Listeria*, *Escherichia*, and *Lactobacillus* strains (Vurma et al., 2006; Malone et al., unpublished data, 2006). Selected strains of *L. monocytogenes* (Scott-A, OSY-8578, and OSY-328) and *E. coli* O157:H7 (EDL-933, MBM-OSY, ASM-OSY) were treated with TBHQ (100 ppm) and pressure (400 MPa, 5 min at 25°C). None of these strains were affected by the TBHQ treatment alone (data not shown). The sensitizing action of TBHQ was observed in both genera, but the enhanced pressure lethality was more pronounced for *E. coli* O157:H7 than for *L. monocytogenes* (Figure 7.1). The presence of TBHQ during HPP increased the lethality of pressure against *Listeria* strains by 2–2.5 logs, whereas this combination treatment enhanced the inactivation *E. coli* O157:H7 strains by > 6 logs. *Lb. plantarum* ATCC 8014 was also tested with this combination treatment, and the barotolerance and the enhanced lethal effects of combination treatment of this strain were comparable to those of the barotolerant *L. monocytogenes* OSY-328 (Figure 7.2).

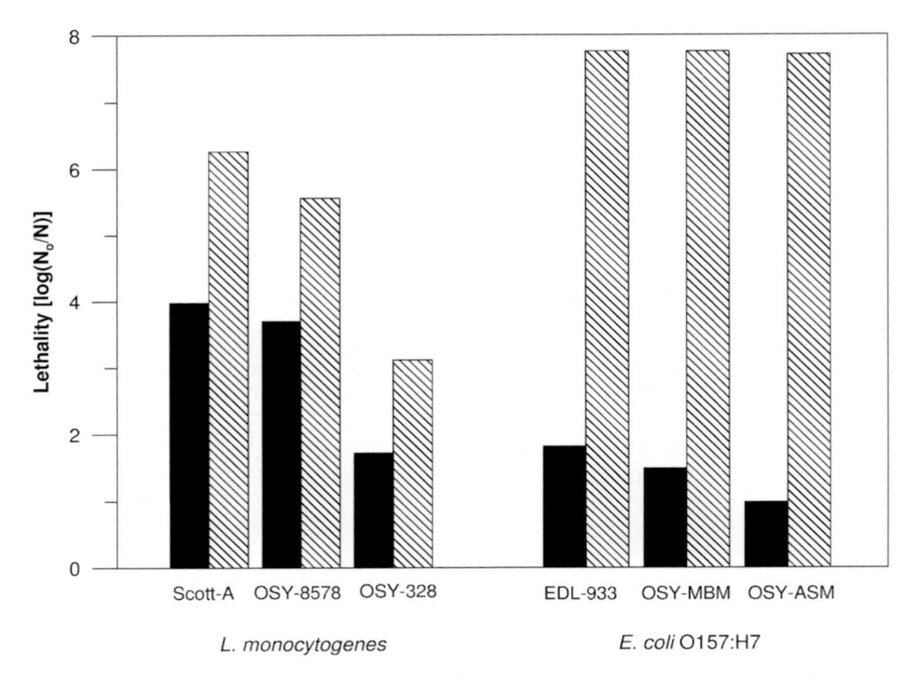

Figure 7.1. Inactivation of different strains of *Listeria monocytogenes* and barotolerant *Escherichia coli* O157:H7 by high pressure at 400 MPa at 25°C for 5 min in the presence of *tert*-butylhydroquinone (100 ppm). Solid bars indicate high pressure only, and hatched bars indicate *tert*-butylhydroquinone and high pressure combination. Data are averages of two to six independent trials. N_0, initial count; N, count after treatment. (Malone et al., unpublished data, 2006; Vurma et al., 2006.)

Inactivation Kinetics of Barotolerant Listeria by Phenolic-Pressure Treatments

Investigating the lethality of different phases of a preservation method (i.e., inactivation kinetics) helps in designing efficient processing technologies without compromising food safety. Recently, inactivation kinetic patterns of different strains of *L. monocytogenes*, in response to TBHQ-pressure combinations, were investigated (Vurma et al., 2006). Pressure come-up time caused some lethality that depended on the bacterial strain and the targeted pressure. The sensitizing action of TBHQ also varied with the strain and the final processing pressure. Results depicting the death behavior of the barotolerant *L. monocytogenes* OSY-328, in the presence of TBHQ, are shown in Figure 7.3. Treating this

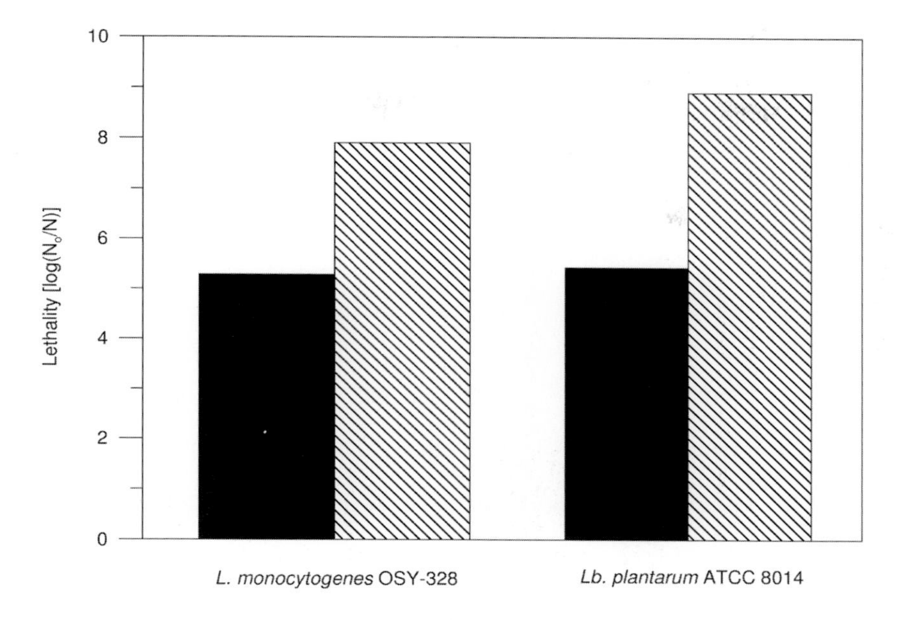

Figure 7.2. Inactivation of barotolerant *Listeria monocytogenes* OSY-328 and *Lactobacillus plantarum* ATCC 8014 by high pressure at 500 MPa at 25°C for 5 min, in the presence of *tert*-butylhydroquinone (100 ppm). Solid bars indicate high pressure only, and hatched bars indicate *tert*-butylhydroquinone and high pressure combination treatment. Data are averages of two to three independent trials. N_0, initial count; N, count after treatment.

strain with pressure resulted in characteristic biphasic inactivation kinetics, that is, a rapid decline in viable count during an initial treatment time followed by a phase of slow or minimal decrease in viability. The second phase (tailing) was more pronounced when HPP was used alone. However, the presence of TBHQ sensitized *L. monocytogenes* to HPP and the tailing behavior was undetectable (Figure 7.3).

Dose-Response of Phenolic-Pressure Treatment against Barotolerant Listeria

Although the lethality of a preservation method against targeted microorganisms generally increases with the intensity of the treatment, the dose-response relationship needs to be assessed before a new processing technology is implemented by food processors. The response

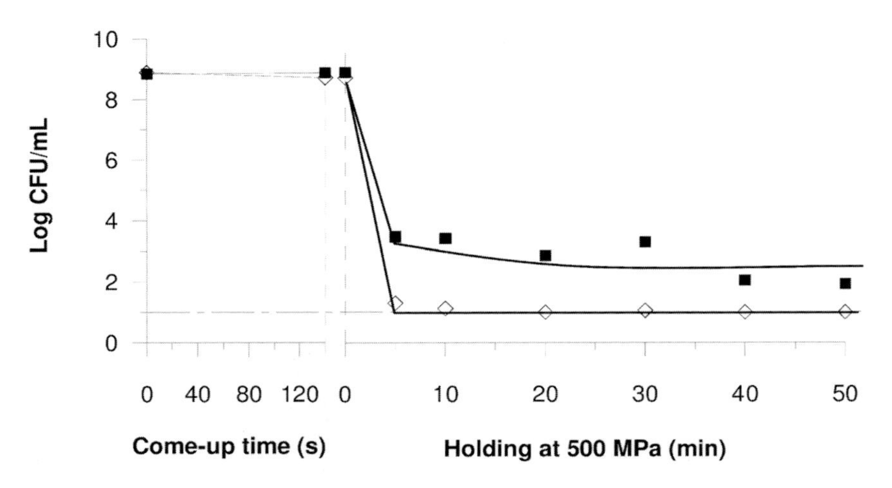

Figure 7.3. Inactivation kinetics of *Listeria monocytogenes* OSY-328 by high pressure treatment (500 MPa, 18–20°C) in the presence of *tert*-butylhydroquinone (100 ppm). ■ high pressure only; ◇ *tert*-butylhydroquinone and high pressure. Data points are averages of three independent trials. The horizontal dashed line indicates the detection limit of analytical method used. Data near or at the dashed line are estimates or below detection limit, respectively. (Adapted from Vurma et al., 2006.)

of the barotolerant *L. monocytogenes* OSY-328 to different phenolic-pressure combinations is shown in Figure 7.4. The enhanced pressure lethality by TBHQ was largest at 500 MPa for this strain. However, this synergistic effect was largest at 400 MPa for *L. monocytogenes* Scott-A and OSY-8578 (data not shown). The investigation also shows that certain threshold pressures should be attained before the sensitization effect of TBHQ becomes apparent. These threshold pressures were < 300 MPa for *L. monocytogenes* Scott-A, and > 300 MPa for *L. monocytogenes* OSY-328.

Mechanism of Sensitizing Microorganisms to High Pressure by Phenolics

Phenolics include a large number of compounds that vary considerably in their properties and their mode of interaction with living cells. A small group of these compounds has been tested in combination with high pressure; among these, a very limited number was investigated

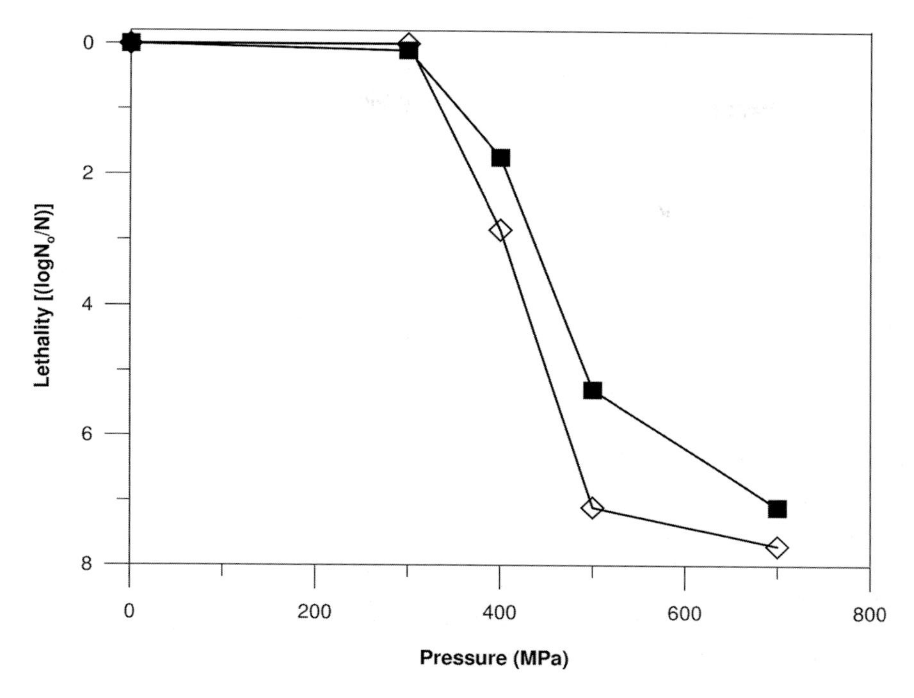

Figure 7.4. Pressure-dose response for *Listeria monocytogenes* OSY-328 at 25°C for 5 min. ■ high pressure only; ◇ *tert*-butylhydroquinone (100 ppm) and high pressure. Data points are averages of two to six independent trials. N_o, initial count; N, count after treatment. (Adapted from Vurma et al., 2006.)

in relation to their ability to sensitize pathogens to HPP. Therefore, covering the mechanism of sensitizing microorganisms to high pressure by phenolics is a difficult task and the proposed mechanisms are highly speculative. Since the authors of this chapter recently investigated some aspects of this mechanism using TBHQ, emphasis will be given to this phenolic compound and related antioxidants.

Phenolic antioxidants are used as food additives to prevent lipid oxidation due to their ability to scavenge oxidative free radicals (Bors and Saran, 1987). Many of the biological actions of phenolics have been attributed to their antioxidant properties through influencing the intracellular redox status (Williams et al., 2004). However, this classical hydrogen-donating antioxidant activity may not explain many of the cellular effects observed when phenolics are combined with HPP. The following discussion highlights cellular mechanisms that are common or complementary between phenolics and HPP. This approach helps in elucidating the mechanism of sensitizing bacteria to HPP by phenolics.

Effects on Cell Membrane

The cytoplasmic membrane of the microbial cell presumably is the target of high pressure (Hoover et al., 1989; Cheftel, 1995). This is probably related to the increase in the melting temperature of membrane lipids by more than 10°C per 100 MPa. Therefore, membrane lipids present in a liquid state at atmospheric pressure will crystallize under high pressure; this may change the structure and permeability of the cell membrane (Cheftel, 1995). The lipophilic phenolic compounds may also target the cytoplasmic membrane, increasing its permeability to ions such as H^+ (Sikkema et al., 1994; Ultee et al., 1998). The hydroxyl groups of phenolics may contribute to the anti-microbial action of these compounds by providing a dissociable H^+ that decreases the pH gradient across the cytoplasmic membrane (Ultee et al., 2002). Therefore, the presence of phenolics may enhance the physical damage to the cytoplasmic membrane during the pressure treatment.

Ultra-high pressure alters the conformation of the bacterial cell envelope (Malone et al., 2002). High pressure also appears to increase the partition of phenolics into cell membranes, or the diffusivity of these compounds through membranes. This hypothesis is supported by previous observations regarding the contribution of pressure to the interaction between water and hydrocarbons (Sawamura et al., 1989) and between water and protein's hydrophobic core (Hummer et al., 1998). Thus, antioxidant phenolics such as TBHQ may partition into the phospholipid areas of the inner membrane in bacterial cells, where many redox reactions are occurring, especially during growth, leading to the production of reactive oxygen species (ROS) and possibly disrupting the energy generating reactions.

Phenolics as Pro-oxidants in vivo

Phenolics such as BHA, TBHQ, or flavonoids are extensively metabolized in vivo, resulting in a significant alteration in their redox potentials (Tajima et al., 1991; Williams et al., 2004). Generally, phenolics undergo intracellular metabolism via conjugation with thiols, particularly glutathione (GSH). The product of this reaction (TBHQ-GSH) has a higher redox potential, compared to TBHQ (Tajima et al., 1991; van Ommen et al., 1992). Furthermore, in vivo changes of TBHQ generate semiquinone radicals through autooxidation, accompanied by the formation of superoxide anion, hydrogen peroxide, and hydroxyl radicals (Kahl et al., 1989; Schilderman et al., 1993).

There is an emerging view supporting the pro-oxidant cytotoxicity of phenolic compounds. Compounds acting as antioxidants to lipid often act as pro-oxidants to other molecules such as DNA and protein (Laughton et al., 1989). Antioxidants capable of reducing Fe^{3+} to Fe^{2+} sometimes act as pro-oxidants. Therefore, a phenolic compound could exhibit antioxidant or pro-oxidant properties depending on the availability of a free radical source (Cao et al., 1997; Li et al., 2000, 2002; Badary et al., 2003; Nemeikaitė-Čėnienė et al., 2005). Li et al. (2002) demonstrated that *tert*-butylquinone (TBQ), semiquinone anion radical, and ROS are formed by Cu^{2+}-dependent TBHQ activation, resulting in DNA damage. Badary et al. (2003) also reported that TBHQ (25–100 μM) caused concentration-dependent DNA damage. According to Kraft et al. (2004), TBHQ is known to induce antioxidant response element (ARE) in mammalian cells, indicating that these cells have induced protective mechanisms against toxicities of electrophiles and ROS. The cellular defense mechanisms against oxidative stress include the production of a reducing agent such as glutathione, transcriptional activation of detoxification enzymes via ARE, and the expression of genes such as OxyR (a bifunctional regulatory protein sensor for oxidative stress) and SoxRS (a transcriptional regulator that senses the redox state of cell). Therefore, phenolic antioxidants such as TBHQ may serve as precursors for DNA-damaging reactive molecules. Conditions that favor the generation of such reactive molecules in bacterial cells may include high pressure treatments.

Involvement of Proteins Containing Iron-Sulfur Clusters

The cellular effects of phenolics will ultimately depend on the extent to which they associate with cells, either by interactions at the membrane or by uptake into the cytosol. Therefore, the combination of membrane disruption, protein destabilization, and intracellular redox stress may all contribute to the synergistic lethality between phenolics and HPP. According to Aertsen et al. (2005), HPP caused intracellular oxidative

Table 7.4. Contribution of *Escherichia coli* genes/operons, which are related to cell's oxidation-reduction reaction and iron homeostasis, to susceptibility to high pressure (HPP) and *tert*-butylhydroquinone (TBHQ) combination (Malone et al., 2006; Malone et al., unpublished data, 2006)

		Relative Susceptibility of Mutants Compared to Wild Types	
Gene/Operon	Function	HPP	HPP + TBHQ
Suf	Iron-sulfur cluster assembly during iron starvation	Resistant	Resistant[a]
Isc	Iron-sulfur cluster biosynthesis, assembly, and regulation	Resistant	Sensitive
Fnr	Transcription regulator for anaerobic respiration	Resistant	Resistant
Nar	Anaerobic respiration, nitrate reductase	No change	Resistant
trxA	Thiol-disulfide redox system	Sensitive	Resistant
Dps	Stress response DNA-binding protein	Sensitive	Sensitive[b]

[a] Significantly more resistant than that treated with HPP alone.
[b] Significantly more sensitive than that treated with HPP alone.

stress and sensitized *E. coli* to the oxidative compounds, plumbagin and *t*-butylhydroperoxide. A recent study (Malone et al., 2006) shows that regulatory protein containing iron-sulfur (Fe-S) clusters and thiol-disulfide redox systems were affected by HPP; this would lead to the disruption of the redox homeostasis in the cell, and impairment of the cell's ability to deal with the intracellular pro-oxidants, generated from TBHQ. Results of a study on *E. coli*, with mutations in relevant genes, provide evidence supporting the former hypothesis (Malone et al., unpublished data. 2006). These mutants were tested for their sensitivities to TBHQ-HPP combination, and results of this study are summarized in Table 7.4.

Iron-sulfur clusters function as key cofactors in diverse cellular processes. Many of the cluster-containing proteins are involved in the electron transfer system (Beinert and Kiley, 1999; Frazzon and Dean, 2003).

In *E. coli*, both *isc* and *suf* operons encode proteins involved in the assembly of Fe-S clusters, and these two systems may be regulated to provide similar functions under distinct conditions. Outten et al. (2004) suggested that *suf* is induced in response to oxidative stress and iron limitation, whereas *isc* induction is related to a disruption of overall Fe-S cluster status and the inability of IscS (cysteine desulfurase, major donor of sulfur atoms for Fe-S cluster), IscU (Fe-S cluster assembly scaffold), and IscA (alternative Fe-S cluster assembly scaffold) to maintain the cluster of the *isc* transcriptional regulator, IscR. Fumarate nitrate reductase (FNR) is one of the most abundant Fe-S proteins in *E. coli,* and it contains an oxygen sensitive Fe-S cluster that controls its DNA binding ability and, thus, gene regulation. Overexposure to oxygen disrupts the $[4Fe\text{-}4S]^{2+}$ complex, producing the cluster-free protein apoFNR [0Fe-0S] (Achebach et al., 2005). Our results indicated that cells capable of producing active Fe-S clusters are sensitive to pressure (Malone et al., 2006; Malone et al., unpublished data, 2006). For example, *E. coli* defective in *suf*, *isc*, and *fnr* were pressure resistant (Table 7.4). In addition, both *suf* and *fnr* mutants were also more resistant, than their wild-type counterparts, to the TBHQ-HPP combination (Table 7.4). These results suggest that HPP targets Fe-S clusters and releases iron ions in the cytosol. If the iron level is high, then *suf* would not be induced regardless of *isc* induction. Iron cations facilitate the Fenton reaction and generate highly reactive hydroxyl radicals that damage cellular DNA (Keyer and Imlay, 1996). The iron cations released in a pressure-treated cell may also exacerbate the oxidation of TBHQ and production of DNA-damaging free radicals, and the presence of TBHQ thus increases the lethality of the pressure treatment.

Nitrate Reductase and Resistance to Phenolics

The transcriptional regulator FNR controls many genes involved in anaerobic metabolism, including those involved in nitrate utilization (*nar* and *nap* operons). Nitrate metabolism involves a highly reactive quinone pool located near the cell periphery (periplasm and cytoplasmic membrane). In *E. coli*, the genes for Nap (periplasmic nitrate reductase) are expressed under nitrate-limiting conditions (Brondijk et al., 2004). The Nar system controls the expression of genes encoding nitrate reductase (Lee et al., 1999). Interestingly, *E. coli* strains with mutations in *nar* operon were more resistant to TBHQ-HPP combination than their wild-type counterparts (Table 7.4). However, there were no significant differences in barotolerance between the isogenic mutant pairs when HPP was used alone. During anaerobic respiration of *E. coli*, nitrate may serve as a terminal electron acceptor from the membrane-bound quinone pool, with nitrate reductase catalyzing these reactions. Buc et al. (1995) found that nitrate reductase, consisting of α and β subunits, can use the artificial electron donor, benzyl viologen, to reduce nitrate. Based on the observations presented in this chapter, it is likely that TBHQ, a highly active redox-cycling compound, also serves as electron donor for the nitrate reductase system. The involvement of TBHQ in the cellular metabolic redox system, acting as a ubiquinone or menaquinone analog, may generate intermediate radicals (phenoxy radicals), and disturb the cell's redox homeostasis or membrane's proton motive force. Since bacterial ubiquinones and menaquinones contain 6–10 isoprenyl units in the third position, and TBHQ has only a *tert*-butyl in the second position, cellular mobility of TBHQ would be much greater than the localized pool of metabolic quinones. Therefore, the presence of TBHQ in cells with nitrate reductase leads to the generation of reactive radicals that cause cell damage or death. Consequently, mutations in genes involving expression of nitrate reductase are beneficial to cells treated with the TBHQ-HPP combination (Table 7.4).

Miscellaneous Mechanisms

Thioredoxin 1 (*trxA*), a primary component of the thiol-disulfide redox system in *E. coli*, is a small electron-transfer protein that can oxidize thiol groups or reduce disulfide bonds of proteins, depending on its redox state, and can also act as a chaperone (Prinz et al., 1997). Thioredoxin 1 is reduced by NADPH in a reaction catalyzed by thioredoxin reductase (*trxB*) and functions in a wide variety of cellular processes, including controlling the cell's redox state and proper protein folding. According to a recent study, the *E. coli trxA* mutant was more sensitive to pressure, but more resistant to the combined TBHQ-HPP treatment than its wild-type counterpart (Malone et al., unpublished data, 2006). It has been reported that a mutant defective in *trxA* creates a high redox potential in the cytosol, and thus leads to constitutive activation of OxyR (Åslund et al., 1999), which would ultimately lead to protection of the cell against oxidative stress, such as that encountered when the TBHQ-HPP combination is applied.

A DNA-binding protein, Dps, is induced by the regulatory protein OxyR in response to H_2O_2 (Altuvia et al., 1994). A *dps* mutant is particularly sensitive to oxidative DNA damage (Martinez and Kolter, 1997). Dps protects the DNA by scavenging free iron, thus protecting the cell against Fenton chemistry (Nair and Finkel, 2004). In vitro, this protein can both store iron and bind to DNA (Almirón et al., 1992). According to recent studies (Malone et al., 2006; Malone et al., unpublished data, 2006), the *dps* mutant was sensitive to pressure, and to the TBHQ-HPP combination, in comparison to its wild-type counterpart. These results further support the hypothesis that free iron ions and/or DNA damage, probably via the Fenton reaction, are related to the TBHQ-assisted high-pressure inactivation.

We investigated whether cell inactivation by the TBHQ-HPP combination can be alleviated by including reducing agents, such as glutathione and cysteine. These reducing agents were added before the TBHQ-HPP treatment of *E. coli* O157:H7 EDL-933. Glutathione and cysteine significantly enhanced the resistance of cells against the TBHQ-HPP treatment (Malone et al., unpublished data, 2006). Glutathione and cysteine are strong reducing agents that have been used to protect cells against oxidative damage or redox stress (Carmel-Harel and Storz, 2000). This suggests that TBHQ obviously increases oxidative damage induced by HPP.

Potential Applications

Some phenolics are approved for use in foods as antioxidants, and their use, in conjunction with HPP, to improve food safety is feasible. Eradication of pathogens, such as *L. monocytogenes* and *E. coli* O157:H7, and spoilage microorganism, such as lactobacilli, may be accomplished by processing ready-to-eat foods with combined phenolic-pressure treatments. The following study illustrates the feasibility of applying TBHQ-HPP, as a post-lethality treatment, to eliminate *L. monocytogenes* in sausage (Chung et al., 2005).

Vienna sausage was inoculated, at 10^6–10^7CFU/g, with *L. monocytogenes* strains that varied considerably in terms of their resistance to processing treatments. Inoculated samples were treated with combinations of HPP (600 MPa at 28°C for 5 min) and TBHQ (100–300 ppm) and the percent positive samples was determined (Table 7.5). All

Table 7.5. Percentage[a] of inoculated sausage sample bags that tested positive for *Listeria monocytogenes* after pressure treatment in the presence of *tert*-butylhydroquinone (adapted from Chung et al., 2005)

Treatments		% Bags Tested *Listeria* Positive[c]	
Pressure (MPa)[b]	TBHQ (ppm)	Scott-A	OSY-328
0	0	100	100
0	100	100	100
0	300	100	100
600	0	85	95
600	100	15	20
600	300	0	0

[a] Means of four trails with five bags each; bags were inoculated to contain 10^6 to 10^7 *Listeria*/g sample.
[b] Pressure: 600 MPa at 28°C for 5 min.
[c] Scott-A is a processing-resistant, and OSY-328 is a processing-sensitive strain.

sample bags, pre-inoculated with the pathogenic strains and treated with TBHQ only, were *Listeria*-positive. The high pressure treatment alone was not effective in eliminating the tested strains and only a modest decrease in the number of positive bags was observed. However, the TBHQ-pressure combination caused a significant decrease in the number of bags testing positive for all strains. When 300 ppm TBHQ was used, *L. monocytogenes* strains were not detected in any of the inoculated sausage bags after treatment with the phenolic-HPP combination. Furthermore, sausage packages treated with the latter combination remained *Listeria*-free for more than a year of storage at both ambient and refrigerated conditions (data not shown). This experiment demonstrates that TBHQ greatly sensitizes foodborne *Listeria* to high pressure, and thus the combination treatment is potentially suitable for commercial applications.

Conclusion

The efficacy of phenolics-HPP combination treatments against dangerous foodborne pathogens were discussed. Selected TBHQ-HPP combinations eliminated the occurrence of "tailing" in the inactivation of pathogen populations; such tailing is evident when high pressure is used as the sole lethal agent. Therefore, TBHQ with HPP may be an effective combination for improving the microbiological safety and quality of foods. The potential use of these combinations in food products, especially in processed meats, is feasible. Possible bacterial inactivation mechanisms by phenolic-HPP combination treatments were discussed, and multiple cellular targets are likely involved in the inactivation of harmful foodborne pathogens. Although studies indicating this synergy against bacterial spores are limited, combining phenolics with HPP is a promising strategy to enhance the safety of food that may have future utility for developing strategies for the elimination of spores from foods.

References

Achebach, S., T. Selmer, and G. Unden. 2005. Properties and significance of apoFNR as a second form of air-inactivated [4Fe-4S] FNR of *Escherichia coli*. *FEBS J.* 272:4260–4269.

Aertsen, A., P. De Spiegeleer, K. Vanoirbeek, M. Lavilla, and C.W. Michiels. 2005. Induction of oxidative stress by high hydrostatic pressure in *Escherichia coli*. *Appl. Environ. Microbiol.* 71:2226–2231.

Al-Khayat, M.A., and G. Blank. 1985. Phenolic spice components sporostatic to *Bacillus subtilis*. *J. Food Sci.* 50:971–974, 980.

Almirón, M., A.J. Link, D. Furlong, and R. Kolter. 1992. A novel DNA-binding protein with regulatory and protective roles in starved *Escherichia coli*. *Genes Dev.* 6:2646–2654.

Alpas, H., N. Kalchayanand, F. Bozoglu, A. Sikes, C.P. Dunne, and B. Ray. 1999. Variation in resistance to hydrostatic pressure among strains of food-borne pathogens. *Appl. Environ. Microbiol.* 65:4248–4251.

Altuvia, S., M. Almiron, G. Huisman, R. Kolter, and G. Storz. 1994. The *dps* promoter is activated by OxyR during growth and by IHF and σ^s in stationary phase. *Mol. Microbiol.* 13:265–272.

Arts, M.J.T.J., G.R.M.M. Haenen, L.C., S.A. Wilms, J.N. Beetsra, C.G.M. Heunen, H.-P. Voss, and A. Bast. 2002. Interactions between flavonoids and proteins: Effect on the total antioxidant capacity. *J. Agric. Food Chem.* 50:1184–1187.

Åslund, F., M. Zheng, J. Beckwith, and G. Storz. 1999. Regulation of the OxyR transcription factor by hydrogen peroxide and the cellular thiol-disulfide stress. *Proc. Natl. Acad. Sci. USA* 96:6161–6165.

Badary, O.A., R.A. Taha, A.M.G. El-Din, and M.H. Abdel-Wahab. 2003. Thymoquinone is a potent superoxide anion scavenger. *Drug Chem. Toxicol.* 26:87–98.

Beinert, H., and P.J. Kiley. 1999. Fe-S proteins in sensing and regulatory functions. *Curr. Opin. Chem. Biol.* 3:152–157.

Bernheim, F. 1972. The effect of chloroform, phenols, alcohols and cyanogens iodide on the swelling of *Pseudomonas aeruginosa* in various salts. *Microbios* 5:143.

Blom, H., E. Nerbrink, R. Dainty, T. Hagtvedt, E. Borch, H. Nissen, and T. Nesbakken. 1997. Addition of 2.5% lactate and 0.25% acetate controls growth of *Listeria monocytogenes* in vacuum-packed, sensory acceptable servelat sausage and cooked ham stored at 4°C. *Int. J. Food Microbiol.* 38:71–76.

Bors, W., and M. Saran. 1987. Radical scavenging by flavonoids antioxidants. *Free Radical Res. Commun.* 2:289–294.

Brondijk, T.C.H., A. Nilavongse, N. Filenko, D.J. Richardson, and J.A. Cole. 2004. NapGH components of the periplasmic nitrate reductase of *Escherichia coli* K-12: Location, topology and physiological roles in quinol oxidation and redox balancing. *Biochem. J.* 379:47–55.

Buc, C.-L. Santini, F. Blasco, R. Giordani, M.L. Cárdenas, M. Chippaux, A. Cornish-Bowden, and G. Giordano. 1995. Kinetic studies of a soluble αβ complex of nitrate reductase A from *Escherichia coli*. Use of various αβ mutants with altered β subunits. *Euro. J. Biochem.* 234:766–772.

Burt, S. 2004. Essential oils: their antibacterial properties and potential applications in foods-a review. *Int. J. Food Microbiol.* 94:223–253.

Cao, G., E. Sofic, and R.L. Prior. 1997. Antioxidant and prooxidant behavior of flavonoids: Structure-activity relationships. *Free Rad. Biol. Med.* 22:749–760.

Carmel-Harel, O., and G. Storz. 2000. Roles of glutathione- and thioredoxin-dependent reduction systems in the *Escherichia coli* and *Saccharomyces cerevisiae* responses to oxidative stress. *Annu. Rev. Microbiol.* 54:439–461.

Chaibi, A., L.H. Ababouch, K. Belasri, S. Boucetta, and F.F. Busta. 1997. Inhibition of germination and vegetative growth of *Bacillus cereus* T and *Clostridium botulinum* 62A spores by essential oils. *Food Microbiol.* 14:161–174.

Chang, H.C., and A.L. Branen. 1975. Antimicrobial effects of butylated hydroxyanisole (BHA). *J. Food Sci.* 40:349.

Cheftel, J.C. 1995. Review: High pressure, microbial inactivation and food preservation. *Food Sci. Technol. Int.* 1:75–90.

Chen, Y., and A.E. Hagerman. 2004. Quantitative examination of oxidized polyphenol-protein complexes. *J. Agric. Food Chem.* 52:6061–6067.

Chung, Y-.K., M. Vurma, E.J. Turek, G.W. Chism, and A.E. Yousef. 2005. Inactivation of barotolerant *Listeria monocytogenes* in sausage by combination of high-pressure processing and food-grade additives. *J. Food Prot.* 68:744–750.

Cléry-Barraud, C., A. Gaubert, P. Masson, and D. Vidal. 2004. Combined effects of high hydrostatic pressure and temperature for inactivation of *Bacillus anthracis* spores. *Appl. Environ. Microbiol.* 70:635–637.

Davidson, P.M. 1997. “Chemical preservatives and natural antimicrobial compounds.” In: *Food Microbiology: Fundamentals and Frontiers,* ed. M.P. Doyle, L.R. Beuchat, and T.J. Montville, pp. 520–556. Washington, DC: ASM, Washington.

———. 2005. “Parabens.” In: *Antimicrobials in Foods*, 3rd ed., ed. P.M. Davidson, J.N. Sofos, and A.L. Sofos, pp. 291–304. Boca Raton, FL: CRC Press.

Davidson, P.M., and A.L. Branen. 1980a. Antimicrobial mechanisms of butylated hydroxyanisole against two *Pseudomonas* species. *J. Food Sci.* 45:1607–1613.

———. 1980b. Inhibition of two psychrotrophic *Pseudomonas* species by butylated hydroxyanisole. *J. Food Sci.* 45:1603–1606.

Davidson, P.M., C.J. Brekke, and A.L. Branen. 1979. Antimicrobial effects of butylated hydroxylanisole (BHA), tertiary butylhydroquinone (TBHQ), and potassium sorbate against *Staphylococcus aureus* and *Salmonella typhimurium*. *Inst. Food Technol. 39th Ann. Meeting*, St. Louis, MO.

Deans, S.G., and G. Ritchie.1987. Antibacterial properties of plant essential oils. *Int. J. Food Microbiol.* 5:165–180.

Denyer, S.P., and W.B. Hugo. 1991. "Biocide-induced damage to the bacterial cytoplasmic membrane." In: *Mechanisms of Action of Chemical Biocides. The Society for Applied Bacteriology, Technical Series* No. 27, ed. S.P. Denyer and W.B. Hugo, pp. 171–188. Oxford, UK: Oxford Blackwell Scientific Publication.

Dorman, H.J.D., and S.G. Deans. 2000. Antimicrobial agents from plants: Antibacterial activity of plant volatile oils. *J. Appl. Microbiol.* 88:308–316.

Draughon, F.A., S.C. Sung, J.R. Mount, and P.M. Davidson. 1982. Effects of parabens with and without nitrite on *Clostridium botulinum* toxin production in canned pork slurry. *J. Food Sci*. 47:1635.

Fogg, A.H., and R.M. Lodge. 1945. The mode of antibacterial action of phenols in relation to drug fastness. *Trans. Faraday Soc*. 41:359.

Frazzon, J., and D.R. Dean. 2003. Formation of iron-sulfur clusters in bacteria: An emerging field in bioinorganic chemistry. *Curr. Opin. Chem. Biol.* 7:166–173.

Furr, J.R., and A.D. Russell. 1972. Some factors influencing the activity of esters of *p*-hydroxybenzoic acid against *Serratia marcescens*. *Microbios* 5:189.

Grant, S., M. Patterson, and D. Ledward. 2000. Food processing gets freshly squeezed. *Chem. Ind*. 24 (January): 55–58.

Gueldner, R.C., D.M. Wilson, and A. Heidt. 1985. Volatile compounds inhibiting *Aspergillus flavus*. *J. Food Chem*. 33:441–443.

Hoover, D.G., C. Metrick, A.M. Papineau, D.F. Farkas, and D. Knorr. 1989. Biological effects of high hydrostatic pressure on food microorganisms. *Food Technol*. 43:99–107.

Hummer, G., S. Garde, A.E. García, M.E. Paulaitis, and L.R. Pratt. 1998. The pressure dependence of hydrophobic interactions is consistent with the observed pressure denaturation of proteins. *Proc Natl. Acad. Sci. USA* 95:1552–1555.

Ismaiel, A.A., and M.D. Pierson. 1990a. Effects of sodium nitrite and origanum oil on growth and toxin production of *Clostridium botulinum* in TYG broth and ground pork. *J. Food Prot*. 53:958–960.

———. 1990b. Inhibition of germination, outgrowth, and vegetative growth of *Clostridium botulinum* 67B by spice oils. *J. Food Prot*. 53:755–758.

Judis, J. 1963. Studies on the mechanism of action of phenolic disinfectants. 2. Patterns of release of radioactivity from *Escherichia coli* labeled by growth on various compounds. *J. Pharm. Sci.* 52:126.

Juneja, V.K., and P.M. Davidson. 1993. Influence of altered fatty acid composition on resistance of *Listeria monocytogenes* to antimicrobials. *J. Food Prot.* 56:302–305.

Jurd, L., A.D. King, K. Mihara, and W.L. Stanley. 1971. Antimicrobial properties of natural phenols and related compounds. 1. Obtusastyrene. *Appl. Microbiol*. 21:507.

Kahl, R., S. Weinke, and H. Kappus. 1989. Production of reactive oxygen species due to metabolic activation of butylated hydroxyanisole. *Toxicology* 59:179–194.

Karatzas, A.K., E.P.W. Kets, E.J. Smid, and M.H.J. Bennik. 2001. The combined action of carvacrol and high hydrostatic pressure on *Listeria monocytogenes* Scott A. *J. Appl. Microbiol.* 90:463–469.

Keyer, K., and J.A. Imlay. 1996. Superoxide accelerates DNA damage by elevating free-iron levels. *Proc. Natl. Acad. Sci. USA* 93:13635–13640.

Klindworth, K.L., P.M. Davidson, C.J. Brekke, and A.L. Branen. 1979. Inhibition of *Clostridium perfringens* by butylated hydroxyanisole. *J. Food Sci.* 44:564–567.

Kraft, A.D., D.A. Johnson, and J.A. Johnson. 2004. Nuclear factor E2-related factor 2-dependent antioxidant response element activation by *tert*-butylhydroquinone and sulforaphane occurring preferentially in astrocytes conditions neurons against oxidative insult. *J. Neurosci.* 24:1101–1112.

Kurita, N., M. Miyaji, R. Kurene, Y. Takahara, and K. Ichimura. 1981. Antifungal activity of components of essential oils. *Agric. Biol. Chem.* 45:945–952.

Lado, B.H., and A.E. Yousef. 2002. Alternative food preservation technologies: Efficacy and mechanisms. *Microbes Infect.* 4:433–440.

Lambert, R.J.W., P.N. Skandamis, P.J. Coote, and G.-J.E. Nychas. 2001. A study of the minimum inhibitory concentration and mode of action of oregano essential oil, thymol and carvacrol. *J. Appl. Microbiol.* 91:453–462.

Laughton, M.J., B. Halliwell, P.J. Evans, and J.R. Hoult. 1989. Antioxidant and pro-oxidant actions of plant phenolics quercetin, gossypol, and myricetin. Effect on lipid peroxidation, hydroxyl radical generation and bleomycin-dependent damage to DNA. *Biochem. Pharmacol.* 38:2859–2865.

Lee, A.I., A. Delgado, and R.P. Gunsalus. 1999. Signal-dependent phosphorylation of the membrane-bound NarX two-component sensor-transmitter protein of *Escherichia coli*: Nitrate elicits a superior anion ligand response compared to nitrite. *J. Bacteriol.* 181:5309–5316.

Li, A.S., B. Bandy, S. Tsang, and A.J. Davidson. 2000. DNA-breaking versus DNA-protecting activity of four phenolic compounds *in vitro*. *Free Rad. Res.* 33:551–566.

Li, Y., A. Seacat, P. Kuppusamy, J.L. Zweier, J.D. Yager, and M.A. Trush. 2002. Copper redox-dependent activation of 2-*tert*-butyl(1,4)hydroquinone: Formation of reactive oxygen species and induction of oxidative DNA damage in isolated DNA and cultures rat hepatocytes. *Mutation Res.* 518:123–133.

López-Malo A., E. Palou, and S.M. Alzamora. 2005. Naturally occurring compounds—Plant sources. In: *Antimicrobials in Foods*, 3rd ed., ed. P.M. Davidson, J.N. Sofos, and A.L. Sofos, pp. 429–452. Boca Raton, FL:CRC Press.

López-Pedemonte, T., A.X. Roig-Sagués, A.J. Trujillo, M. Capellas, and B. Guamis. 2003. Inactivation of spores of *Bacillus cereus* in cheese by high hydrostatic pressure with the addition of nisin or lysozyme. *J. Dairy Sci.* 86:3075–3081.

Mackey, B.M., K. Forestière, and N. Isaacs. 1995. Factors affecting the resistance of *Listeria monocytogenes* to high hydrostatic pressure. *Food Biotechnol.* 9:1–11.

Malone, A.S., Y.-K. Chung, and A.E. Yousef. 2006. Genes of *Escherichia coli* O157:H7 involved in high pressure resistance. *Appl. Environ. Microbiol.* 72:2661–2671.

Malone, A.S., Y.-K. Chung, and A.E. Yousef, unpublished results

Malone, A.S., T.H. Shellhammer, and P.D. Courtney. 2002. Effects of high pressure on the viability, morphology, lysis, and cell wall hydrolase activity of *Lactococcus lactis* subsp. *cremoris*. *Appl. Environ. Microbiol.* 68:4357–4363.

Martinez, A., and R. Kolter. 1997. Protection of DNA during oxidative stress by the nonspecific DNA-binding proteins Dps. *J. Bacteriol.* 179:5188–5194.

Nair, S., and S.E. Finkel. 2004. Dps protect cells against multiple stress during stationary phase. *J. Bacteriol.* 186:4192–4198.

Nakayama, A., Y. Yano, S. Kobayashi, M. Ishikawa, and K. Sakai. 1996. Comparison of pressure resistance of spores of six *Bacillus* strains with their heat resistance. *Appl. Environ. Microbiol.* 2:3897–3900.

Nemeikaitė-Čėnienė, A., A. Imbrasaitė, E. Sergedienė, and N. Čėnas. 2005. Quantitative structure-activity relationships in prooxidant cytotoxicity of polyphenols: Role of potential of phenoxyl radical/phenol redox couple. *Arch. Biochem. Biophys.* 441:182–190.

Outten, F.W., O. Djaman, and G. Storz. 2004. A *suf* operon requirement for Fe-S cluster assembly during iron starvation in *Escherichia coli*. *Mol. Microbiol.* 52:861–872.

Papadopoulou, A., and R.A. Frazier. 2004. Characterization of protein-polyphenol interactions. *Trends Food Sci. Technol.* 15:186–190.

Prinz, W.A., F. Åslund, A. Holmgren, and J. Beckwith. 1997. The role of the thioredoxin and glutaredoxine pathways in reducing protein disulfide bonds in the *Escherichia coli* cytoplasm. *J. Biol. Chem.* 272:15661–15667.

Reddy, N.R., M.D. Pierson, and R.V. Lechowich. 1982. Inhibition of *Clostridium botulinum* by antioxidants, phenols, and related compounds. *Appl. Environ. Microbiol.* 43:835–839.

Rico-Munoz, E., E.E. Bargiota, and P.M. Davidson. 1987. Effects of selected phenolic compounds on the membrane-bound adenosine triphosphatase of *Staphylococcus aureus*. *Food Microbiol.* 4:239.

Robach, M.C., L.A. Smoot, and M.D. Pierson. 1977. Inhibition of *Vibrio parahaemolyticus* 04:K11 by butylated hydroxyanisole. *J. Food Prot.* 40:549–555.

Shahidi, F., P.K. Janitha, and P.D. Wanasundara. 1992. Phenolic antioxidants. *Crit. Rev. Food Sci. Nutri.* 32:67–103.

Sakanaka, S., L.R. Juneja, and M. Taniguchi. 2000. Antimicrobial effects of green tea polyphenols on thermophilic spore-forming bacteria. *J. Biosci. Bioeng.* 90:81–85.

Sawamura, S., K. Kitamura, and Y. Taniguchi. 1989. Effect of pressure on the solubilities of benzene and alkylbenzenes in water. *J. Phys. Chem.* 93:4931–4935.

Schilderman, P.A.E.L., J.M.S. van Maanen, F.J. ten Vaarwerk, M.V.M. Lafleur, E.J. Westmijze, F. ten Hoor, and J.C.S. Kleinjans. 1993. The role of prostaglandin H synthase-mediated metabolism in the induction of oxidative DNA damage by BHA metabolites. *Carcinogenesis* 14:1297–1302.

Shearer, A.E.H., C.P. Dunne, A. Sikes, and D.G. Hoover. 2000. Bacterial spore inhibition and inactivation in foods by pressure, chemical preservatives, and mild heat. *J. Food Prot.* 63:1503–1510.

Sierra, G. 1970. Inhibition of the amino acid induced initiation and germination of bacterial spores by chlorocresol. *Can. J. Microbiol.* 16:51.

Sikkema, J., J.A.M. De Bont, and B. Poolman. 1994. Interactions of cyclic hydrocarbons with biological membranes.*J. Biol. Chem.* 269:8022–8028.

———. 1995. Mechanisms of membrane toxicity of hydrocarbons. *Microbiol. Rev.* 59:201–222.

Sofos, J.N., and F.F. Busta. 1980. Alternatives to the use of nitrite as an antibotulinal agent. *Food Technol.* 34:244–251.

Soni, M.G., S.L. Taylor, and N.A. Greenberg. 2002. Evaluation of the health aspects of methyl paraben; a review of the published literature. *Food Chem. Toxicol.* 40:1335–1373.

Tajima, K., M. Hashizaki, K. Yamamoto, and T. Mizutani. 1991. Identification and structure characterization of S-containing metabolites of 3-*tert*-butyl-4-hydroxyanisole in rat urine and liver microsomes. *Drug. Metab. Dispos.* 19:1028–1033.

Tay, A., T.H. Shellhammer, A.E. Yousef, and G.W. Chism. 2003. Pressure death and tailing behavior of *Listeria monocytogenes* strains having different barotolerances. *J. Food Prot.* 66:2057–2061.

Tsuchiya, H. 1999. Effects of green tea catechins on membrane fluidity. *Pharmacology* 59:34–44.

Ultee, A., M.H.J. Bennik, and R. Moezelaar. 2002. The phenolic hydroxyl group of carvacrol is essential for action against the food-borne pathogen *Bacillus cereus. Appl. Environ. Microbiol.* 68:1561–1568.

Ultee, A., L.G.M. Gorris, and E.J. Smid. 1998. Bactericidal activity of carvacrol towards the food-borne pathogen *Bacillus cereus. J. Appl. Microbiol.* 85:211–218.

Ultee, A., E.P.W. Kets, and E.J. Smid. 1999. Mechanisms of action of carvacrol on the food-borne pathogen *Bacillus cereus. Appl. Environ. Microbiol.* 65:4606–4610.

Ultee, A., R.A. Slump, G. Steging, and E.J. Smid. 2000. Antimicrobial activity of carvacrol toward *Bacillus cereus* on rice. *J. Food Prot.* 63:620–624.

Van Ommen, B., A. Koster, H. Verhagen, and P.J. van Bladeren. 1992. The glutathione conjugates of *tert*-butylhydroquinone as potent redox cycling agents and possible reactive agents underlying the toxicity of butylated hydroxyanisole. *Biochem. Biophys. Res. Comm.* 189:309–314.

Vurma, M., Y.-K. Chung, T.H. Shellhammer, E.J. Turek, and A.E. Yousef. 2006. Use of phenolic compounds for sensitizing *Listeria monocytogenes* to high-pressure processing. *Int. J. Food Microbiol.* 106:269–275.

Williams, R.J., J.P.E. Spencer, and C. Rice-Evans. 2004. Flavonoids: Antioxidants or signaling molecules? *Free Rad. Biol. Med.* 36:838–849.

Yousef, A.E., R.J. Gajewski II, and E.H. Marth. 1991. Kinetics of growth and inhibition of *Listeria monocytogenes* in the presence of antioxidant food additives. *J. Food Sci.* 56:10–13.

Yuste, J., R. Pla, M. Capellas, and M. Mor-Mur. 2002. Application of high-pressure processing and nisin to mechanically recovered poultry meat for microbial decontamination. *Food Control* 13:451–455.

Chapter 8

Functional Genomics for Optimal Microbiological Stability of Processed Food Products

Stanley Brul, Hans van der Spek, Bart J.F. Keijser, Frank H.J. Schuren, Suus J.C.M. Oomes, and Roy C. Montijn

Introduction

The presence and growth of spoilage and pathogenic microorganisms is a key concern to the food processing industry. Some of the most difficult structures to deal with are the extremely heat resistant bacterial endospores. Spores create problems due to their ability to survive classical thermal treatments. Then, after surviving such treatments, the spores are able to repair damage and, in many cases, subsequently germinate and form regular vegetative cells. Research on food spoilage *Bacillus subtilis* isolates using the Amplified Fragment Length Polymorphism (AFLP) technology and micro-arrays has identified a number of genome sequences that correlate to the spore's level of heat resistance. Strains could be classified according to these DNA markers. In addition, it was shown with the sequenced *B. subtilis* laboratory strain that sporulation in the presence of a complex mix of minerals, prominently including calcium, promotes the thermal resistance of developing spores. This physiological observation correlated with increased expression during sporulation of genes encoding small acid-soluble spore proteins (SASPs).

Screening food ingredients using the DNA-chip based techniques identifying the above indicated molecular markers is foreseen in the (near) future for identification of spoilage and pathogenic microorganisms as well as prediction of their preservation stress resistance in response to lethal treatments (thermal, high-pressure, etc.). Currently various projects aiming at the integration of genomics data and micro(nano)-technology, a prerequisite if the alluded-to ingredient quality control is going to succeed, are running and new ones are being set up. Information from these projects will be used together with the requirements of product organoleptic quality to derive robust integrated food safety and food quality processing parameters. Such parameters should form the basis of developing models for future food quality assurance systems. Concepts discussed in this chapter have been adapted from Brul et al. (2005a, 2006) and Marthi et al. (2004).

Background

The food processing industry is faced with an ever-increasing demand for safe and minimally processed wholesome foods. In order to come to a knowledge-based understanding for achieving these goals, rather than a mainly empirical approach of combining appropriate preservation hurdles, many groups are currently exploring the use of the booming technology of "genomics" in setting food processing parameters. While this book focuses primarily on the application of high pressure processing (HPP) to inactivate bacterial spores, we would like to illustrate the potential of applying genomics strategies to the study of the behavior of bacterial sporeformers, using thermal processes as an example that might one day be applicable to HPP. The data is most completely available for thermal processes, and, in any case, the genomics approach does not change if thermal processing is replaced by HPP (only the actual data will). As a start, conceptually all "genomics" tools are summarized in Figure 8.1.

To fully exploit the power of these novel technologies in the area of food microbiology, a new approach to science is currently being developed. For the ensemble of these "functional genomics" studies the term "systems biology" has been coined. This approach is well illustrated in Westerhoff (2005), including references therein and other papers in that special issue of *Current Opinion in Biotechnology.* During the past decades a reductionist approach has governed most of the research in life science. The disadvantage of this classical approach is that scientists, prior to performing an experiment, decide to focus on only one

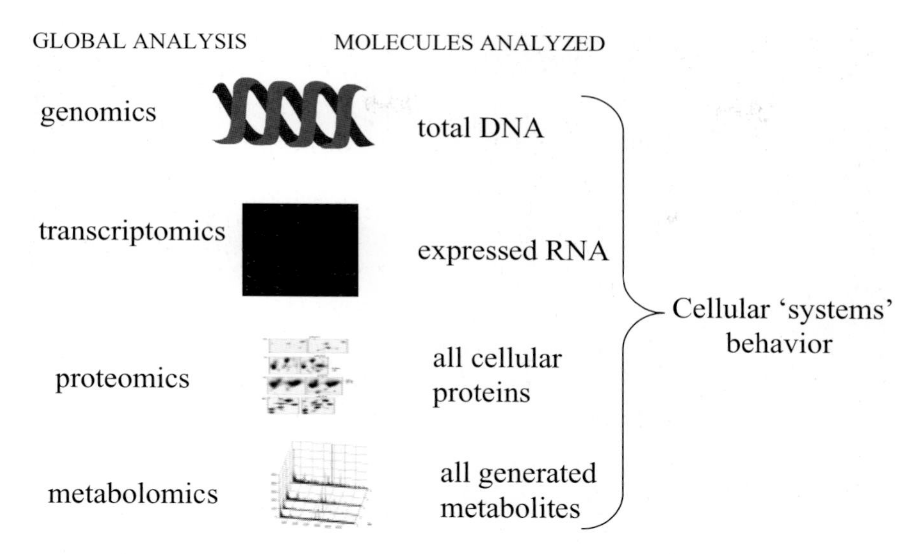

Figure 8.1. A schematic view of the level at which the various genomics technologies operate and contribute to the understanding of cellular behavior. Classical genomics is concerned with the sequencing and annotation of the full genome of a cell. Transcriptomics deals with the study of the global gene expression using whole genome micro-arrays (Nouwens et al., 2000; Lucchini et al., 2001). Proteomics is the study of the full cellular complement of proteins using gel-electrophoresis and mass-spectrometry techniques, while metabolomics is now emerging for the study of all generated cellular metabolites (see, e.g., Weimer and Mills, 2002, and Bruin and Jongen, 2003, for discussions on applications in food science).

aspect, based on the available, and often by its nature limited, experimental evidence for the mechanisms governing microbial behavior mixed with what is at best "expert knowledge." To open the "black box" of cellular behavior, molecular biology has proven to be instrumental (see Figure 8.2).

Developments in that area started with the pioneering work of Mendel on genetics and have accelerated at an increasingly more rapid pace with the discovery of the double-helix structure of DNA in the early 1950s. The sequencing of the first (microbial) genomes comprising a few mega base pairs in the 1990s has paved the way for the large efforts in organism sequencing as we know them today. The human genome consisting of some 3.2 giga base pairs has ("as good as completely") been unraveled with "only" a few hundred gaps to go (see, e.g., Eichler et al., 2004). Developments in sequencing optimization are now such

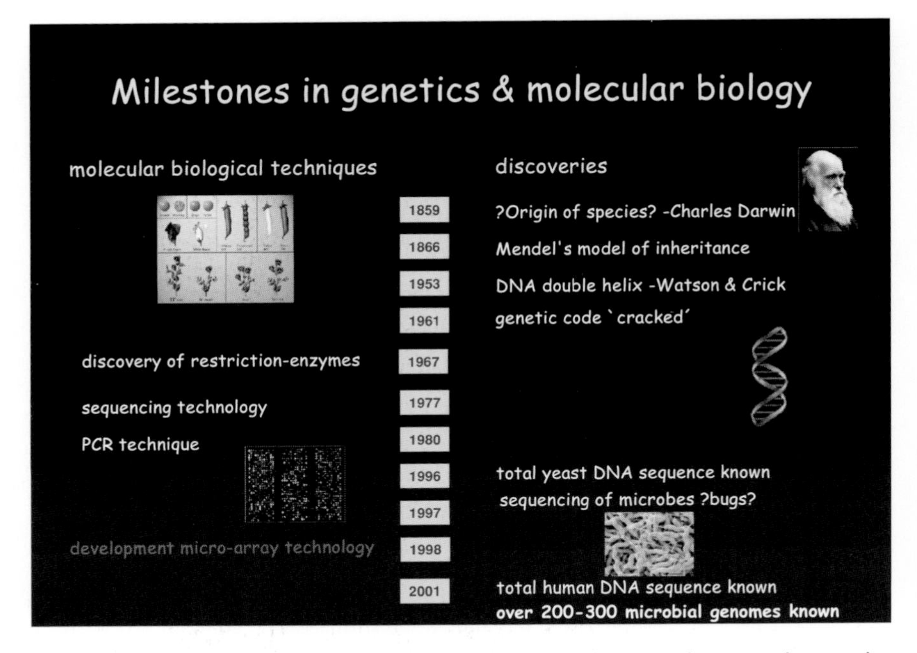

Figure 8.2. Molecular biology in a historical perspective. Developments in genetics and molecular biology are presented from the nineteenth century until the most recent events.

that we can reach a potential speed of something like 300,000 parallel sequencing reactions, making it in principle feasible to (draft) sequence a microbe in several hours rather than days (Margulies et al., 2005). This is very relevant in validating our extensive knowledge of what one might call "pet" lab-strains against real wild-type pathogenic or spoilage isolates.

The novel "genomics" technologies make use of a holistic, unbiased approach, in which the whole set of cellular biomolecules is studied under the relevant experimental conditions and in a relevant timeframe. Generally, genome and genome usage (transcript "ome") information is still the most completely available, since, technically speaking, this is easiest to standardize and automate. It should be realized, though, that by far, not all regulation occurs at the level of gene expression. Thus, the analysis of the presence and activity of proteins is, in fact, as important, if not even more important, to understanding the physiological behavior of cells in the context of their genes and gene expression. Proteomics

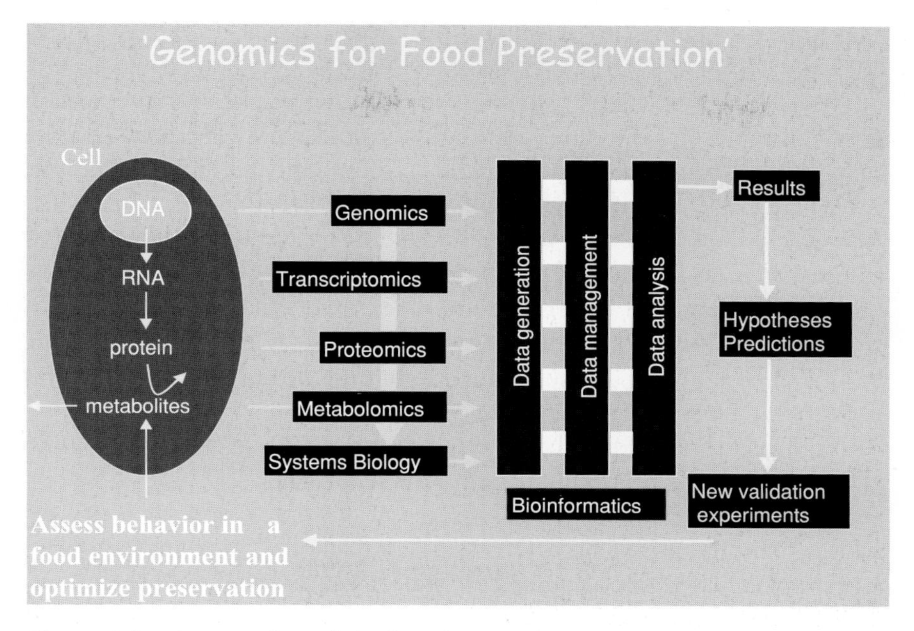

Figure 8.3. An overview of the iterative experimental and computational cycles that are operative in functional genomics and systems biology research (discussed in Westerhoff and Palsson, 2004). The main challenge is to store and interpret the data coming from transcript and proteome analysis using biological knowledge and modern bioinformatics to handle such large data sets. The aim is to come to accurate predictions of the mode of action on microorganisms of preservation processes in combination with product composition that can feed process optimization both at the onset of the definition of new processes and at the optimization of process routines.

approaches are fortunately becoming more and more potent and quantitative, which will undoubtedly facilitate this understanding (Kolkman et al., 2005; Phillips and Bogyo, 2005).

The application of bioinformatics to analyze genome-wide transcript (gene-expression) data (e.g., through pattern recognition routines) currently is standard and results in identification by way of a correlative analysis of specific biomarkers relevant to the conditions under study. If desired, detailed studies on selected biomarker molecules can be started at this point to further analyze in detail the mode of action of antimicrobial treatments using more and more systems analysis approaches (see below and Figure 8.3).

Genomics in Modelling Microbes in Food

An important area where genomics is expected to play a major role in food microbiology is in guiding predictive modelling characterizing the behavior of spoilage and pathogenic microorganisms during and after processing. The analyses will provide a molecular fingerprint, a molecular mechanistic basis, for the survival strategies of microorganisms of interest in various foods. This not only creates a robust basis for models of microbial behavior under the food processing and distribution conditions studied, but also will allow (certain levels of) extrapolation of the results obtained with one set of food manufacturing/preservation stress conditions to another (Brul et al., 2002, 2006).

As a result, we should be able to predict which of the preservation treatments might be optimally combined in order to come to an effective, possibly synergistic, inhibition of unwanted microbial growth while ensuring optimal product organoleptic quality. In addition, a more (mechanistic) approach applied to food preservation should lead to increased robustness of predictions with reference to microbial food stability (Abee and Wouters, 1999; Brul et al., 2003; van Schaik and Abee, 2005).

A Case Study of *Bacilli* Forming Highly Thermoresistant Endospores

In *Bacillus subtilis* we have recently studied the genome-wide response of cells toward environmental conditions in terms of forming extremely high versus normal heat resistant spores, as measured according to the method of Kooiman (Kooiman, 1973; see also Cazemier et al., 2001). Spore formation in *Bacillus* is a highly regulated process, the basis of which is outlined briefly here.

Sporulation is a dramatic morphological differentiation process, leading to the retreat of the chromosome into a small heat resistant capsule, where it will stay in an inactive dormant state until the environmental conditions become favorable for supporting growth. Sporulation is tightly regulated and is only initiated when alternative survival strategies are no longer effective (Errington, 2003; Barak et al., 2005; Stephenson and Lewis, 2005). In addition, the efficiency of sporulation depends strongly on such conditions as the composition of the medium and other factors.

The sporulation process itself can be subdivided into a sequence of different stages, based on specific morphological changes that occur. These stages are characterized by specific groups of genes that are activated during each sporulation stage. The most significant changes in gene regulation during sporulation are brought about by the appearance of new sigma factors, which confer new promoter specificities on RNA polymerase, thereby activating new groups of genes. Figure 8.4 gives an overview of the individual cell differentiation stages that are generally distinguished during sporulation and discussed below (for a comprehensive review see Hilbert and Piggot, 2004).

During vegetative growth most *B. subtilis* genes are transcribed by RNA polymerase containing the major sigma factor, σ^A, while some genes are transcribed by RNA polymerase containing the minor sigma factor, σ^H. In response to starvation (i.e., nutrient depletion), σ^A and σ^H direct transcription of genes whose products relocate septum formation from the mid-cell to a polar position, and partition one copy of the chromosome to the larger mother cell and the other copy to the smaller forespore (discussed in Barak and Wilkinson, 2005). Shortly after the septum forms, σ^F becomes active in the forespore and, shortly thereafter, σ^E activity is observed in the mother cell. Products of both σ^F- and σ^E-controlled genes are required for the phagocytic-like process of engulfment, which results in the forespore being surrounded by two membranes.

The factor σ^G becomes active in the engulfed forespore, in which it directs transcription of genes encoding proteins that condense and protect the chromosome and prepare the spore for the germination process. Shortly after σ^G becomes active, σ^K becomes active in the mother cell. Both σ^E and σ^K are needed for synthesis of the cortex and coat layers that encase the forespore. The cortex is a loosely cross-linked peptidoglycan formed between the membranes surrounding the forespore, largely by products of σ^E-controlled genes. Coat proteins assemble on the forespore surface, and most are transcribed by σ^K-RNA polymerase. σ^K probably also directs the transcription of genes involved in lysis of the mother cell and release of the mature spore.

Interestingly, when a growing *B. subtilis* culture is observed in time it will become apparent that the timing of sporulation differs substantially between individual cells, despite the fact that cells are genetically

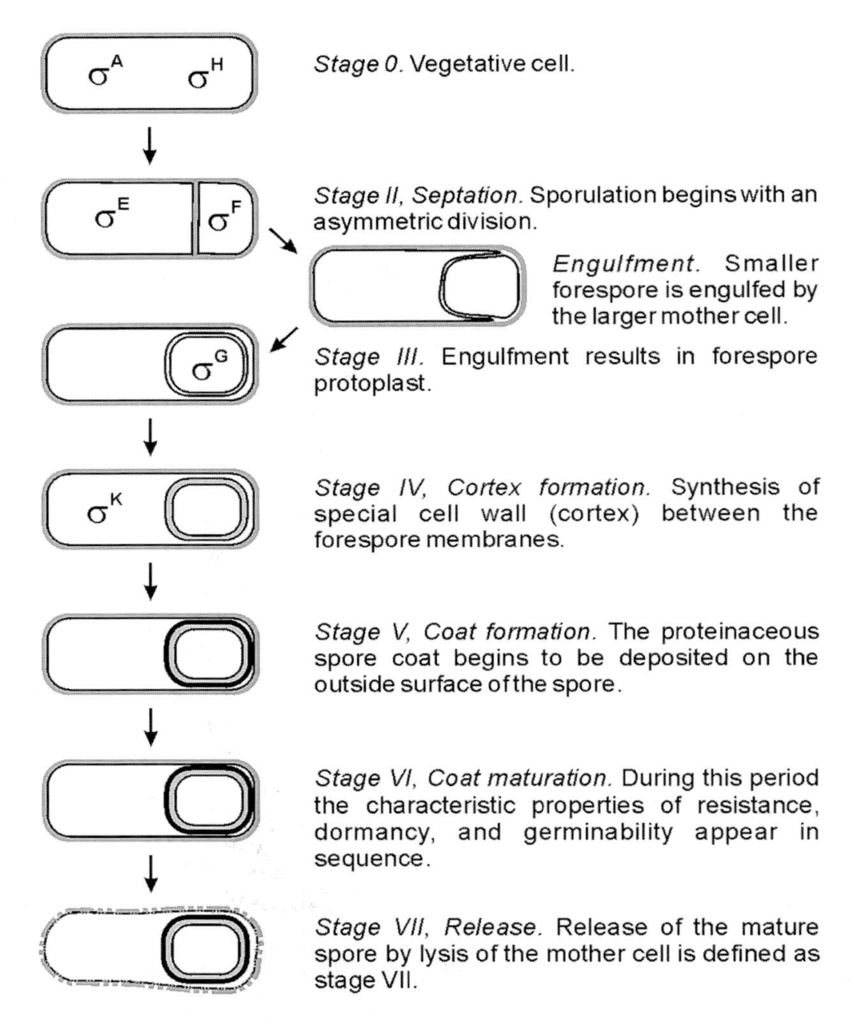

Figure 8.4. Sporulation as a differentiation process for *Bacillus subtilis*. The figure depicts the stages of sporulation of *Bacilli* (omitting *Stage I. Axial filament formation*) with the known regulatory sigma factors (see, e.g., Errington, 2003).

homogeneous and are subjected to the same environmental conditions. There are cells that sporulate relatively early, others will sporulate late, and there are cells that do not sporulate at all. The mechanism behind the heterogeneous development of spores is currently being uncovered. Veening et al. (2005) have shown that bistability at the level of the central regulator Spo0A-phosphate is at least one component at its basis. Heterogeneity is also observed in the stress resistance of the resulting spores (see, e.g., Kort et al., 2005, for an account of the heterogeneity of spore thermal stress resistance).

It is likely that the heterogeneous responses of spore populations to thermal stress and during germination originate with the development of the spore. From a transcriptome analysis it became clear that during sporulation various genes were expressed preferentially in cells sporulated in the test condition, a medium containing a relatively high level of minerals. Minerals present during sporulation, such as calcium, potassium, iron, manganese, and magnesium, are known to lead to the formation of extreme heat resistant spores (see Cazemier et al., 2001, and for a full discussion of the transcriptome data Oomes and Brul, 2004). The data suggested a functional role for the α-β type small, acid-soluble spore proteins (SASPs) in mediating spore heat resistance. This was further corroborated by independent earlier experiments by other researchers who showed that spores obtained from mutant *Bacillus subtilis* cells unable to produce SASPs were significantly more heat sensitive (Setlow et al., 2000).

Comparing gene expression of the lab-strain PS832 during sporulation of the cells with the food spoilage isolate, strain A163, indicated that in the sporulating cells of the respective strains, different sets of genes are activated. Interestingly, irrespective of the sporulation conditions *Bacillus subtilis* strain A163 is always prone to generate spores that have a higher endogeneous thermal resistance (Oomes and Brul, 2004). Figure 8.5 shows a typical example of such a comparison of these two strains in early stage sporulating cells. The data show the occurrence of (i) the large numbers of "y" genes (genes with unknown functionality), and (ii) the SASP, a gene in the group of differentially expressed genes (for further details see Oomes and Brul, 2004). Additionally, it has been shown recently in our laboratory that spores of strain A163 can lose their extreme thermal resistance, and that this loss correlates with a change in the level of cross-reactivity with an antiserum containing antibodies against among others, SASP of Class A (O'Brien et al., unpublished results).

Currently, these experimental approaches complemented with novel proteomics approaches are being applied to other strains of *Bacilli* isolated from products or ingredients used in food manufacturing. We focus here on *Bacillus subtilis* and *Bacillus sporothermodurans*.

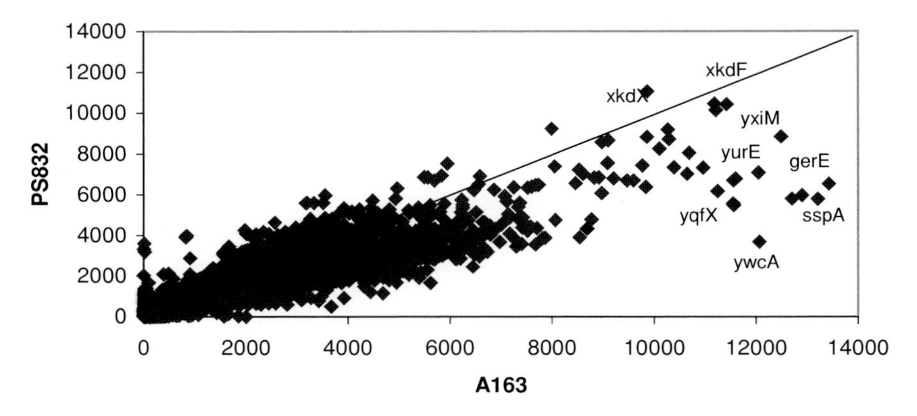

Figure 8.5. A comparison of gene expression (transcriptomics) of *Bacillus subtilis* A163 (product isolate) versus *Bacillus subtilis* PS832. The cells were grown in nutrient broth containing a mineral mix (MM) of calcium, potassium, iron, manganese, and magnesium. Subsequently, they were harvested at various stages into sporulation and total RNA was extracted. The (m)RNA from laboratory strain PS832 was labelled with the red fluorescent dye Cy5 and the (m)RNA from the food spoilage *Bacillus subtilis* isolate A163 was labelled with the green fluorescent dye Cy3. The RNA samples were next mixed and hybridized on a micro-array slide containing essentially all open-reading frames of *Bacillus subtilis.* The axes give absolute fluorescence intensities as they were read from the hybridization scanner. The comparison shown is for cells at the early to middle sporulation stage (taken from Oomes and Brul, 2004. Further details are available in the "Materials and Methods" section of that paper).

These spoilage isolates are generally identified by means of fatty acid analysis and/or 16S rDNA and DNA/DNA hybridization (Kort et al., 2005). Current findings are that the presence of calcium and possibly magnesium during spore formation is key in the development of extreme heat resistance of these spores, presumably, at least in part, mediated through differential gene expression during sporulation (Oomes and Brul, 2004). The heat resistance properties are given in Table 8.1 for a treatment of 3 min at 111–121°C, from spores that were cultivated in a defined liquid medium (MOPS buffered) and in a complex liquid medium (nutrient broth) supplemented with a mixture of mineral salts as described. Clearly these two (extreme) sporulation conditions lead to significantly different levels of heat resistance of the resulting spores.

DNA micro-array analysis, analogous to the analysis used in the studies by Oomes and Brul (2004), is now being used to analyze the various

Table 8.1. Different levels of heat inactivation (expressed as log kills) of spores obtained from different types of media[a]

Strain	Origin	Temp. (°C)	Inactivation (liquid defined media without added minerals)[b]	Inactivation (nutrient broth supplemented with minerals)[b]
B. subtilis PS832 (wt168 trp+)	Culture collection	111[c]	4.5	6.4
B. subtilis A163	Processed food	121	4.4	−0.6 (!)
B. subtilis CC16	Processed food	115	1.6	0.9
B. subtilis IIC14	Ingredient	121	6.0	3.2
B. subtilis CC2	Processed food	121	≥7.1	1.7
B. sporo-thermodurans IC4	Processed food	121	2.9	−0.7 (!)

[a] Strains isolated from food spoilage sources were cultured and sporulated. The heat resistance of the resulting spores was assessed in the incubation of spores suspended in Tryptic Soy broth (TSB) at a treatment time of 3 min at the temperatures indicated. Measurements were done according to the Kooiman (1973) method. Thermal inactivation resistance varied widely among the individual strains and conditions, thermal resistance being increased in spores derived from complex media (nutrient broth; NB). Strain PS 832 is a prototrophic revertant of strain 168 BGSC code 1A700. Furthermore, most noticeably increased mineral content in the sporulation medium (primarily the increase in calcium levels) led to increased thermal resistance. The differences in log kill are due to the different initial inoculation levels used and not to true differences in inactivation. More technical details are given in Oomes and Brul (2004) and Kort et al. (2005). Adapted from Brul et al. (2006). See Oomes et al. (2007) for further details.
[b] The minerals consisted of a mixture of $MgSO_4$, KCl, $CaCl_2$, $MnSO_4$, and $FeSO_4$ (Oomes and Brul, 2004).
[c] These incubations showed a nearly full inactivation.
(!) Indicates heat activated spore germination.

gene-expression profiles of cells sporulating under these conditions. The results will be analyzed again for correlations between the up (or down) regulation of gene expression and the development of high thermal stress resistance. Similar approaches will be applicable to other preservation techniques such as high pressure processing. The generation of predictive tests for strains of sporeforming *Bacilli* with a high propensity to the formation of preservation resistant spores is being pursued (see, e.g., Brul et al., 2002, and Keijser, 2007).

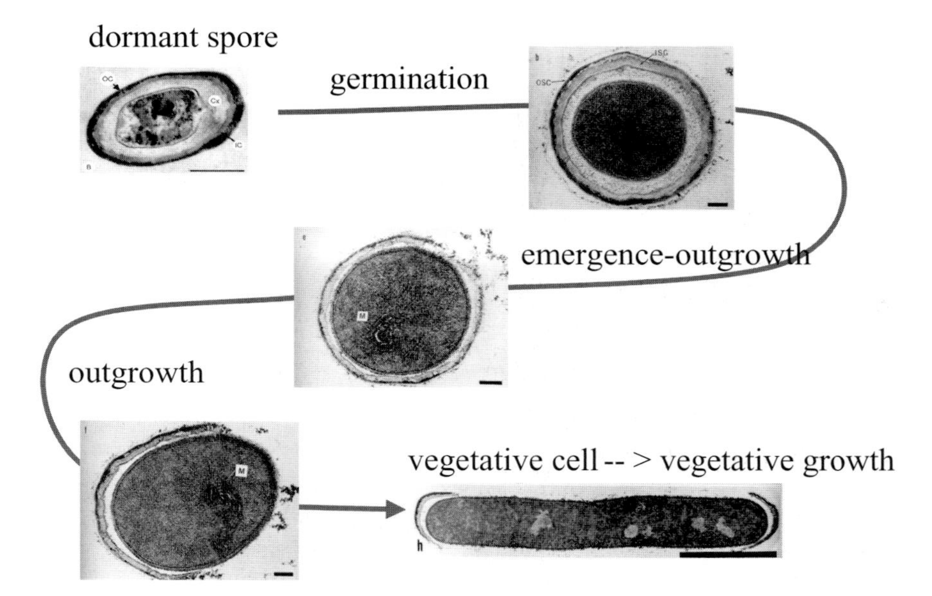

Figure 8.6. Various germination stages of *Bacillus subtilis* spores. The bars in the germination emergence/outgrowth and outgrowth phases indicate a size of 0.1 micrometer. The bar in the vegetative cell indicates 0.5 micrometer.

Aside from characterizing the inherent thermal resistance of microorganisms, the technique of genome-wide transcript analysis also allows for a detailed time resolved assessment of the events during spore germination (Brul et al., 2005a; Keijser et al., 2007). Briefly, bacterial spores first germinate under optimal conditions to get a reference data set. Figure 8.6 describes the various stages of spore germination that were analyzed, and Figure 8.7 gives a summary of the results obtained. We concluded that spore germination proceeds through a highly dynamic sequence of specific morphologic and coupled gene-expression events. Transcription was initiated 5–10 minutes after the onset of germination, as characterized by the excretion of dipicolinic acid, water uptake, and consequent swelling of the spores.

Interestingly, we observed the specific temporal activation of DNA repair genes as the germination process progressed. The repair/recovery mechanisms activated during the process included genes encoding proteases involved in recycling unfolded proteins, DNA repair genes, DNA topoisomerases, RNA modification genes, and genes encoding proteins

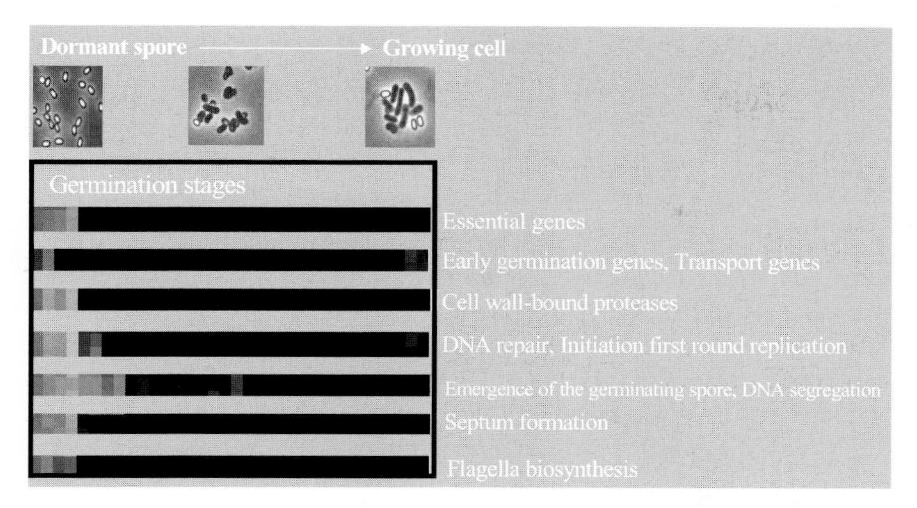

Figure 8.7. Application of genome-wide transcript analysis to germinating *Bacillus subtilis* spores. Schematic representation of transcriptional events during the transition of a dormant *B. subtilis* spore to an actively growing vegetative cell. Top figures show phase contrast images of morphological changes of the germinating and outgrowing spore. The images in the lower part of the figure show the expression of genes during the specific developmental stages in spore germination and outgrowth. Dark shades of grey indicate (increased) presence and a light shade absence of transcripts. RNA was isolated from germinating spores essentially as described by Oomes and Brul (2004) and modified by Keijser et al. (2007).

key to cortex lysis during germination. The data suggest that spores have an inherent "safety valve" to ensure that they repair any damage they may incur before they proceed with outgrowth to becoming a vegetative cell.

Early data of various groups pointed to the fact that in order to let UV-damaged spores repair their DNA properly, some level of germination is necessary (see, e.g., Fajardo-Cavazos et al., 1993; Setlow and Setlow, 1996). An analysis of various mutants in early germination genes is in progress. Under optimal conditions, all of the mutants still germinate, albeit some of them exhibit clearly perturbed kinetics. Currently, we are assessing what these mutants do under acid stress germination conditions and what the behavior of thermally injured spores is. The data indicate that the behavior of certain biomolecules in spores may be used as predictive markers for the level of thermal injury to spores (Keijser, 2007). In all instances the analysis of the large amounts of data poses a major challenge, similar to the situation with genomics data in any other field of microbiology.

Genomics Data Analysis

Primary data analysis involves grouping genes that are induced or repressed in response to a given treatment and is the topic of many studies. Unsupervised data analysis is possible using principal component analysis techniques. To illustrate this, we turn our attention to the use of genomics to predict microbial behavior upon exposure of vegetative cells to elevated temperatures. These results have been described previously (Brul et al., 2006). Briefly, target microorganisms were grown in a medium mimicking conditions commonly found in food matrices and subjected to various heat treatments. Immediately following the application of heat stress to microorganisms, RNA molecules of the microorganisms were isolated and used for transcriptome analysis. Expression patterns relating various treatment conditions were generated and analyzed by principal component analysis. The effects of different temperatures on growth could be clearly distinguished on the basis of this analysis.

An alternative approach has been worked out by Boorsma et al. (2005), who have developed a method to score the difference between the mean expression level of predefined groups of genes (e.g., regulated by common transcription factors) and that of all other genes on a microarray. These categories can be scored without the need to apply cut-offs to the expression level of individual genes (see, for an example of its use, Lascaris et al., 2003). In this way immediately after analysis of a (stress) response profile some assumptions on the underlying mechanistic basis of the cellular response may be made. Figure 8.8 illustrates the principle of this approach.

In all cases, transcript, protein, and metabolic data must be structured in large databases where they serve as a reference guide for further study. The power of genomics then lies in the combination of a transcriptome (completeness) and proteome/metabolome (level of relevance) analysis. It is evident that a full understanding of cellular (eco)physiology will depend on a proper integration of the ecological information (which microbial species/strains are present), molecular data, the stress response options that microbial cells have, and the available food product (related) substrate. We are fortunately now at a stage where this

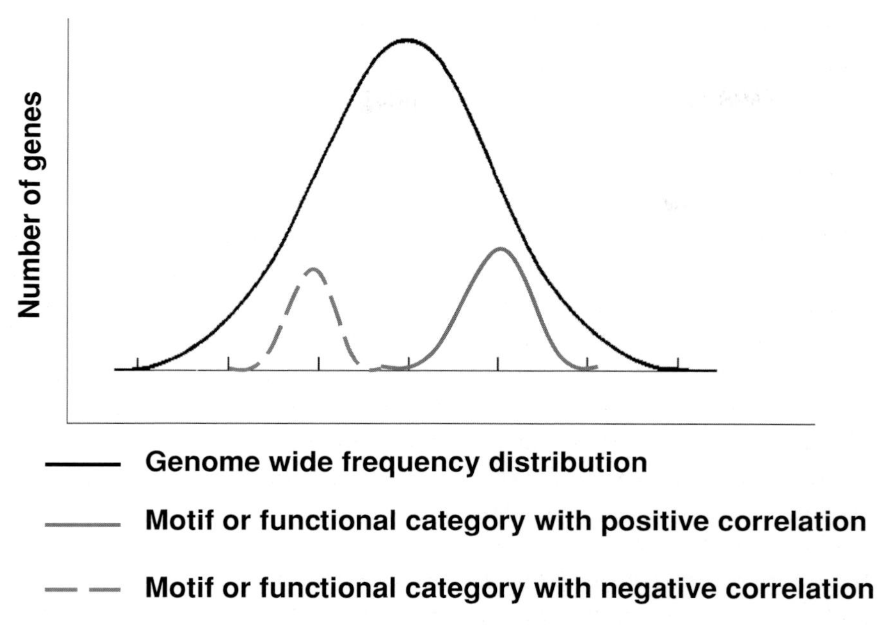

Figure 8.8. The principle of REDUCE (Regulator Element Detection Using Correlation with Expression; see Bussemaker et al., 2001, and also Boorsma et al., 2005, for a further extension of the tool), which uses a statistical approach to find motifs based on (a limited set of) micro-array experiments. By applying an unbiased search, REDUCE selects those sequence motifs whose presence in upstream regions of a gene correlates with an induction or repression of gene expression. Regions of up to 600 base pairs located on the coding strand of the genome are generally used for analysis. Next to continuous motifs also motifs with spacing of up to 20 base pairs may be used in the analysis. Sequence motifs may also originate from direct biochemical binding studies in, for example, Chip-Chip studies (see, e.g., Boorsma et al., 2005).

approach is emerging. Preservation stress resistance and/or virulence of pathogenic microorganisms are just two of the physiological phenomena accurately documented in more and more databases (discussed extensively in, e.g., Brul et al., 2003, and Brul et al., 2005b).

Applying Genomics in the Total Food Chain

Microbial genomics can be applied in various areas of the total food chain. In many cases, these applications involve establishing a liaison with technology developers in areas such as micro- and nanotechnology. At the recent European Congress on Biotechnology in Denmark (2005), this was exemplified by presentations on the development of innovative quality control systems for fermented food, food processing ingredients, and the process lines themselves (Pedersen et al., 2005; see also the congress issue of the *Journal of Biotechnology*). Next to these possible in-factory applications there are already laboratory tools available based on genome sequences that allow for detailed pathogen diagnosis (see, e.g., Dobrindt et al., 2003).

The integration of molecular biology and classical physiology into what is now called "systems biology" is in our view the major challenge and driving force for all of the mentioned genomics applications (see, e.g., Palsson, 2006), and also those in the context of the total food chain. The understanding of cellular responses at the level of molecular events opens up the way to integrally assess at the molecular level microbial responses to environmental conditions as they are found in ingredients; during food processing, storage, and transportation; and finally upon arrival at the retailer and purchase by the consumer. This will allow for the development of mechanistic growth and inactivation models with a truly predictive power. Klipp et al. (2005) and Milo et al. (2002), respectively, give examples of possibilities for highly detailed models for the development of environmental stress resistance of *Saccharomyces cerevisiae* and *Escherichia coli*. In addition, models describing the kinetics of the intracellular metabolic processes with their linked regulatory systems have been described (see, e.g., Brul and Klis, 1999; Teusink, 1999; Stephanopoulos and Kelleher, 2001; Yang et al., 2005; Mensonides et al., 2007; Mensonides, unpublished results). All of these efforts converge in the booming field of systems biology (reviewed in Stelling, 2004; Westerhoff and Palsson, 2004).

It should be realized that the optimal use of the mechanistic knowledge of microbial behavior in the context of the total food chain can be made only if this data can be put in the context of organoleptic product parameters, such as taste flavor, texture, nutritional value, and so forth. Predictive food quality assurance models based on insight into the underlying mechanisms of microbial behavior will (have to) make use of physical parameters pertaining to the heat capacity and heat conductivity of compounds, heat transfer, and mass transfer occurring during processing, and biochemical and nutritional parameters pertaining to the food. It is these types of multi-dimensional models postulated by Bruin and Jongen (2003) in their monumental review of 25 years of food processing that will allow us to reach an optimal balance between efficient product processing, product safety, and product quality.

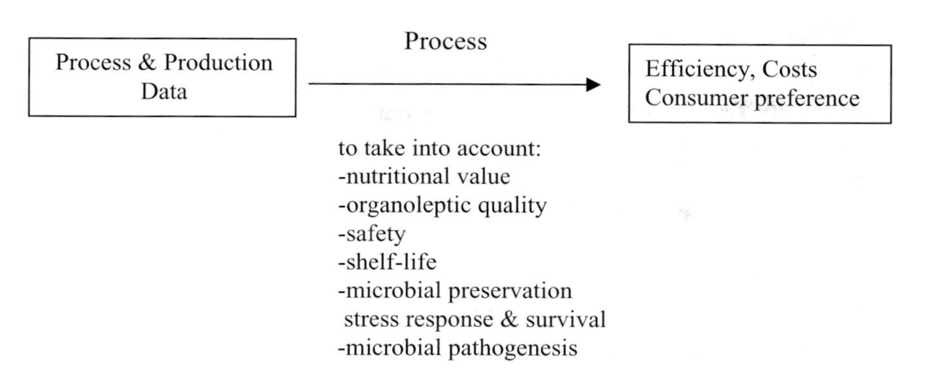

Figure 8.9. A schematic representation of the key events to be considered in food processing. Here physical, chemical, and (micro)biological aspects of product processing need to be studied in an integrated manner to come to desired product attributes (adapted from Jongen, 2001; see also Bruin and Jongen, 2003).

Figure 8.9 summarizes these views in an integrated process design concept (see also Jongen, 2001, and Verrips et al., 2001). Finally, the current scientific developments also facilitate the construction of tools that allow for tracing and tracking of microorganisms (quality control) throughout the whole food chain based on their physiological and molecular ecological characteristics.

In conclusion, genomics projects have been and are continuing to be defined in order to obtain a transparent view of the food chain with respect to the presence of microbes associated with food spoilage/food safety issues. These projects are joint efforts by academia, research institutes, and major multi-national food processing companies. It will be through such collaborative public-private consortia that we expect to see further progress in general and more in particular in unravelling the structure and behavior of bacterial sporeformers. With this data we should then be able to simultaneously feed newly developed detection technologies that will allow isolation, monitoring, and control in the food chain of these, as coined by some, ultimate survival capsules of microorganisms.

Further Reading

A useful URL regarding (functional) genomics and in particular the newly developing and highly relevant field of systems biology is http://www.systemsbiology.org. Applications of a "systems" approach (foreseen) in the food industry including primarily food microbiology are discussed in the document describing a new European technology platform for food research that can be downloaded at http://etp.ciaa.be.

More current literature in the field of predictive (micro)biological modelling of cellular behavior may be obtained from Westerhoff (2005) and elsewhere in the special issue on systems biology of *Current Opinion in Biotechnology* edited by Westerhoff (vol. 16). Finally, in spring 2007 the book *Modelling Microorganisms in Food* was published (edited by Brul, van Gerwen, and Zwietering). Experts in various fields of growth, inactivation, and "lag" model development pertaining to foods contributed to the book.

Acknowledgments

The authors would like to thank all collaborators at Unilever and Amsterdam University in sharpening their thoughts. In particular we would like to thank Dr. Balkumar Marthi, Science Area Leader, Advanced Food Microbiology at the Unilever Food and Health Research Institute.

References

Abee, T., and J.A. Wouters. 1999. Microbial stress response in minimal processing. *Int. J. Food Microbiol.* 40:65–91.

Barak, I., E. Ricca, and S.M. Cutting. 2005. From fundamental studies on sporulation to applied spore research. *Mol. Microb.* 55:330–338.

Barak, I., and A.J. Wilkinson. 2005. Where asymmetry in gene expression originates. *Mol. Microbiol.* 57:611–620.

Boorsma, A., B.C. Foat, D. Vis, F. Klis, and H.J. Bussemaker. 2005. T-profiler: Scoring the activity of predefined groups of genes using gene expression data. *Nucl. Acid Res.* 33:592–595.

Bruin, S., and T. Jongen. 2003. Food process engineering: The last 25 years and challenges ahead. *Comprehen. Revs. in Food Sci. and Food Safety* 2:42–81.

Brul, S., B.J.F. Keijser, H. Van der Spek, S.J.C.M. Oomes, and R. Montijn. 2005a. Genomics applications in food preservation and safety research.*Agro. Food Industry High-Tech* 16:34–36.

Brul, S., and F.M. Klis. 1999. Mechanistic and mathematical inactivation studies of food spoilage fungi. Fungal Genet Biol. 27:199–208

Brul, S., F.M. Klis, D. Knorr, T. Abee, and S. Notermans. 2003. "Food preservation and the development of microbial resistance; past present and future." In: *Food Preservation Techniques,* ed. P. Zeuthen and L. Bogh-Sorensen, pp. 526–552. Cambridge, UK: Woodhead.

Brul, S., F.M. Klis, S.J.C.M. Oomes, R.C. Montijn, F.H.J. Schuren, P. Coote, and K.J. Hellingwerf. 2002. Detailed process design based on genomics of survivors of food preservation processes. *Trends Food Sci. Technol.* 13:325–333.

Brul, S., F. Schuren, R. Montijn, B.J.F. Keijser, H. Van der Spek, and S.J.C.M. Oomes. 2006. The impact of functional genomics on microbiological food quality and food safety. *Int. J. Food Micro.* 112:195–199.

Brul, S., S. Van Gerwen, and M. Zwietering (eds.). 2007. *Modelling Microorganisms in Food*, ed. Cambridge, UK: Woodhead.

Brul, S., J. Wells, and J. Ueckert. 2005b. "Understanding the behaviour of pathogens in the food chain; Food production animals, Food preservation treatment, survival and resistance development." In: *Understanding Pathogen Behaviour,* ed. M. Griffiths, pp. 390–410. Cambridge, UK: Woodhead.

Bussemaker, H.J., H. Li, and E.D. Siggia. 2001 Regulatory element detection using correlation with expression. *Nature Genetics* 27(2):167–171.

Cazemier, A.E., S. Wagenaars, and P.F. Ter Steeg. 2001. Effect of sporulation and recovery medium on the heat resistance and amount of injury of spoilage Bacilli. *J. Appl. Microbiol.* 90:761–770.

Dobrindt, U., F. Agerer, K. Michaelis, A. Janka, C. Buchrieser, M. Samuelson, C. Svanborg, G. Gottschalk, H. Karch, and J. Hacker. 2003. Analysis of genome plasticity in pathogenic and Commensal *Escherichia coli* isolates by use of DNA arrays. *J. Bact.* 185:1831–1840.

Eichler, E.E., R.A. Clark, and X. She. 2004. An assessment of the sequence gaps: Unfinished business in a finished human genome. *Nature Rev. Genet.* 5:345–354.

Errington, J. 2003. Regulation of endospore formation in *Bacillus subtilis. Nat. Rev. Microbiol.* 1:117–126.

Fajardo-Cavazos, P., C. Salazar, and W.L. Nicholson. 1993. Molecular cloning and characterization of the *Bacillus subtilis* spore photoproduct lyase (spl) gene, which is involved in repair of UV radiation-induced DNA damage during spore germination. *J. Bacteriol.* 175:1735–1744.

Hilbert, D.W., and P.J. Piggot. 2004. Compartmentalisation of gene expression during *Bacillus subtilis* spore formation. *Microbiol. Mol. Biol. Rev.* 68:234–262.

Jongen, T. 2001. e-ntegrated design: Where product, process and consumer come together. *Unilever Research Prize lecture* 2000, Vlaardingen, The Netherlands.

Keijser, B.J.F. 2007. Control of Preservation by Biomarkers EP 05077246.6.

Keijser, B.J.F., A. ter Beek, H. Rauwerda, F. Schuren, F. Montijn, R. van der Spek, H., and S. Brul. 2007. Analysis of temporal gene expression during Bacillus subtilis spore germination and outgrowth. J. Bacteriol. 189:3624–3634.

Klipp, E., B. Nordlander, R. Kruger, P. Gennemark, and S. Hohmann. 2005. Integrative model of the response of yeast to osmotic shock. *Nature Biotechnol.* 23:975–982.

Kolkman, A., E.H. Dirksen, M. Slijper, and A.J. Heck. 2005. Double standards in quantitative proteomics: Direct comparative assessment of difference in gelelectrophoresis and metabolic stable isotope labeling. *Mol. Cell Proteomics* 4:255–266.

Kooiman, W.J. 1973. "The screw cap tube technique: A new and accurate technique for the determination of the wet heat resistance of bacterial spores." In: *Spore Research*, ed. A.N. Barker, G.W. Gould, and J. Wolf, pp. 87–92. London: Academic Press.

Kort, R.A., A.C. O'Brien, I.H. Van Stokkum, S.J.C.M. Oomes, W. Crielaard, K.J. Hellingwerf, and S. Brul. 2005. Assessment of heat resistance of bacterial spores from food product isolates by fluorescence monitoring of dipicolinic acid release. *Appl. Environ. Microbiol.* 71:3556–3564.

Lascaris, R., H.J. Bussemaker, A. Boorsma, M. Piper, H. Van der Spek, L. Grivell, and J. Blom. 2003. Hap4p overexpression in glucose-grown Saccharomyces cerevisiae induces cells to enter a novel metabolic state. *Genome Biol.* 4:R3.

Lucchini, S., A. Thompson, and J.C.D. Hinton. 2001, 31 July. Microarrays for microbiologists. *Microbiology* 147:1403–1414.

Margulies, M., et al. 2005. Genome sequencing in microfabricated high-density picolitre reactors. *Nature Biotechnology,* published online.

Marthi, B., E.V. Vaughan, and S. Brul. 2004. Functional genomics and food safety. *New Food* 7:14–18.

Mensonides, F. 2007. How Saccharomyces cerevisiae copes with heat stress; an experimental and theoretical study. Doctoral thesis, University of Amsterdam.

Mensonides, F., B. Bakker, S. Brul, K.J. Hellingwerf, and J.M. Teixeira de Mattos. 2007. "A kinetic model as a tool to understand the response of *Saccharomyces cerevisiae* to heat exposure." In: *Modelling microorganisms in food*, ed. S. Brul, S. van Gerwen, and M. Zwietering, pp. 228–249. Cambridge (UK), Woodhead.

Milo, R., S. Shen-Orr, S. Itzkovitz, N. Kashtan, D. Chlovskii, and U. Alon. 2002. Network motifs: Simple building blocks of complex networks. *Science* 298:824–827.

Nouwens, A.S., S.J. Cordwell, M.R. Larsen, M.P. Molloy, M. Gillings, M.D. Wilcox, and B.J. Walsh. 2000. Complementing genomics with proteomics: The membrane subproteome of *Pseudomonas aeruginosa* PAO1. *Electrophoresis* 21:3797–3809.

O'Brien, A., R. Kort, E. Willems, and S. Brul, *unpublished results*.

Oomes, S.J.C.M., and S. Brul. 2004. The effect of metal ions commonly present in food on gene expression of sporulating *Bacillus subtilis* cells in relation to spore wet heat resistance. *Innov. Food Sci. Emerg. Technol.* 5:307–316.

Oomes, S.J.C.M., A.C.M. van Zuijlen, J.O. Hehenkamp, H. Witsenboer, J.M.B.M. van der Vossen, and S. Brul. 2007. The characterization of *Bacillus* spores occurring in the manufacturing of (low acid) canned products. *Int. J. Food Microbiol.* in press.

Palsson, B.O. (ed.). 2006. *Systems Biology, Properties of Reconstructed Networks.* Cambridge, UK: Cambridge University Press.

Pedersen, L.H., J. Nielson, and S. Gammeltoft (eds.). 2005. ECB-12 abstracts. Bringing genomes to life. *J. of Biotechnol.* 118(1), suppl. 1:1–190.

Phillips, C.I., and M. Bogyo. 2005. Proteomics meets microbiology: Technical advances in global mapping of protein expression and function. *Cell. Microbiol.* 7:1061–1076.

Setlow, B., K.A. McGinnis, K. Rakousi, and P. Setlow. 2000. Effects of major spore-specific DNA binding proteins on *Bacillus subtilis* sporulation and spore properties. *J. Bacteriol.* 182:6906–6912.

Setlow, B., and P. Setlow. 1996. Role of DNA repair in *Bacillus subtilis* spore resistance. *J. Bacteriol.* 178:3486–3495.

Stelling, J. 2004. Mathematical models in microbial systems biology. *Curr. Opin. Microbiol.* 7:513–518.

Stephanopoulos, G., and J. Kelleher. 2001. How to make a superior cell. *Science* 292:2024–2025.

Stephenson, K., and R.J. Lewis. 2005. Molecular insights into the initiation of sporulation in Gram-positive bacteria: New technologies for an old phenomenon. *FEMS Microbiol. Rev.* 281–301.

Teusink, B. 1999. Exposing a complex metabolic system glycolysis in *Saccharomyces cerevisiae*. PhD thesis, University of Amsterdam.

Van Schaik, W., and T. Abee. 2005. The role of sigmaB in the stress response of Gram-positive bacteria—targets for food preservation and safety. *Curr. Opin. Biotechnol.* 16:218–224.

Veening, J.W., L.W. Hamoen, and O.P. Kuipers. 2005. Phosphatases modulate the bistable sporulation gene expression pattern in *Bacillus subtilis*. *Mol. Microbiol.* 56:1481–1494.

Verrips, C.T., M.M.C.G. Warmoeskerken, and J.A. Post. 2001. General introduction to the importance of genomics in food biotechnology and nutrition. *Curr. Opinion Biotechnol.* 12:483–487.

Weimer, B., and D. Mills. 2002. Enhancing foods with functional genomics. *Food Technology* 56:184–189.

Westerhoff, H.V. 2005. Systems biology . . . in action. *Curr. Opin. Biotechnol.* 16:326–328.

Westerhoff, H.V., and B.O. Palsson. 2004. The evolution of molecular biology into systems biology. *Nature Biotechnology* 22:1249–1252.

Yang, C.R., B.E. Shapiro, S.P. Hung, E.D. Mjolsness, and G.W. Hatfield. 2005. A mathematical model for the branched chain amino acid biosynthetic pathways of *Escherichia coli* K12. *J. Biol. Chem.* 280:11224–11232.

Chapter 9

Determination of Quality Differences in Low-Acid Foods Sterilized by High Pressure versus Retorting

---

Ming H. Lau and Evan J. Turek

Introduction

High pressure processing (HPP) is an emerging food preservation technology that uses very high pressures and short times to inactivate spoilage microorganisms and pathogens. HPP can be used for two purposes: pasteurization or sterilization.

In high pressure pasteurization, pressure is used to inactivate vegetative microorganisms to produce shelf-stable high-acid foods or to extend the shelf life of refrigerated low-acid foods. In HP pasteurization, relatively little heat is used to slightly elevate processing temperatures to 20–45°C (Table 9.1). HP pasteurization therefore has the advantage of inducing fewer changes to product characteristics (flavor, aroma, texture, color, and nutrition) as compared to heat pasteurization (Hayashi, 1989; Mertens, 1992; Knorr, 1993; Galazka and Ledward, 1995; Thakur and Nelson, 1998; Yen and Lin, 1996, 1999; Kimura et al., 1994; Parish, 1998). Currently, there are several companies in Canada, France, Japan, Spain, and the United States now using high pressure to pasteurize commercial products for ensuring food safety and extending product shelf life (Table 9.2).

In contrast to high pressure pasteurization, high pressure sterilization employs both pressure and heat synergistically to inactivate vegetative microorganisms *and* spores resulting in shelf-stable low-acid foods

Table 9.1. Comparison of processing conditions for high pressure pasteurization and high pressure sterilization

	Pressure MPa (kpsi)	Time, Min	Temperature, °C
Pasteurization	500–600 (73–87)	5–10	initial: 0–42 at pressure: 20–45
Sterilization	700–1000 (102–145)	2–5 single- or multi-pulses	initial: 70–90 at pressure: ≥106

Table 9.2. Commercial high pressure pasteurized products

Company Name	Location	Products
Avomex Inc./Fresherized Foods	United States	Guacamole, salsa, meal kits, fruit smoothies, juices
Calavo Growers, Inc.	United States	Guacamole
Campofrio Alimentacion	Spain	*RTE meat products
Esteban Espuña, S.A.	Spain	*RTE meat products
Hannah International	United States	Hummus, dips, spreads
Hormel Foods Corp.	United States	*RTE meat products
Joey Oysters Inc.	United States	Oysters
Leahy Orchards Inc.	Canada	Apple sauce/fruit blends
Lovitt Farms Inc.	United States	Apple cider
Meidi-ya	Japan	Jams and fruit toppings
Motivatit Seafoods, Inc.	United States	Oysters
Pampryl	France	Citrus juices
Perdue Farms, Inc.	United States	*RTE chicken

* RTE = ready-to-eat.

(Table 9.1). This process is still considered experimental, although several patents have been granted that teach methods for achieving commercial sterility (Wilson and Baker, 2000, 2001; Meyer, 2000, 2001). In these patents, various permutations of high pressures, pressure cycling, temperature, and treatment times have been claimed to inactivate bacterial spores. The types of spores used in the examples were limited to non-pathogenic species and the most pressure-resistant spores have yet to be rigorously identified. Additional information may likely be required before regulatory approvals are granted to market products treated in these manners commercially.

To date, only a few experimental units exist that can operate under the conditions commonly believed to be needed to sterilize low-acid foods (i.e., $\geq$700 MPa and $\geq$105°C). The units capable of generating these conditions of high pressures and temperatures typically have small internal volumes. In general, these units are suitable for microbiological experiments, but they are also typically too small to produce sufficiently large quantities of food needed for carrying out sensory evaluations. In 2001, a consortium led by the US Army Natick Soldier Research, Development, & Engineering Center (Natick, Massachusetts) brought together a number of industrial partners to further the development of HP sterilization under a Dual-Use Science and Technology (DUST) initiative. The DUST program convened experts from commercial industry, government agencies, and academia with the common purpose of achieving marketable products made commercially sterile through HPP. A 35 L vessel was constructed by Avure Technologies (Kent, Washington) suitable for conducting both quality and microbiological studies.

Determining the best methods for achieving the conditions needed to ensure the safety of HP-sterilized foods is currently the subject of intense, active academic investigation. Presently, the objective of this study is to critically assess HP-treated foods and establish the quality benefits that this HP process might afford the consumer. Following the teachings of Meyer (with permission) in US Patent 6,177,115, a variety of foodstuffs were prepared and processed under HP conditions that were reported to result in sterility. For comparison, the identical products were prepared by traditional thermal retorting. To maximize the quality of the retorted samples, these were prepared in pouches to minimize the retorting times.

Materials and Methods

Product Selection

The five products selected for evaluation were chicken breast, salmon filet, egg omelet, fried potato wedges, and green beans. All of the product formulations were intentionally kept "simple," with the objective of focusing the assessment on discerning intrinsic differences in product quality that could be attributed directly to the differences associated with the high pressure and retort processes. Some pre-testing of the products was done to gain experience with high pressure sterilization as well as thermal processing, and product preparation procedures and processing conditions were adjusted to allow for a fair and meaningful comparison of the two processes.

Product Preparation

Chicken

Fresh chicken breasts were sprinkled with a dry BBQ spice rub (ingredients: salt, spices, and sugar) and placed on a hot grill pan for 30 seconds on each side to sear grill marks.

Salmon

Fresh salmon fillets were immersed in brine (7.5% salt and 11% sugar) for 1 hr to improve the flavor and texture. Results of pre-testing indicated that salmon has a rubbery texture when subjected to HP sterilization conditions, if the brining process is eliminated. After equilibrating in the brine solution for 1 hr, each salmon fillet was patted dry with paper towels and sprinkled with a dry lemon/black pepper spice blend.

Egg

Fresh whole eggs were whipped in a blender while also adding 0.2% citric acid (to prevent "greening"), 23% canola oil (to soften texture), and 0.2% xanthan gum (to prevent syneresis). Results of pre-testing showed that HP eggs tend to have a rubbery texture, which could only be ameliorated by the addition of oil to act as a plasticizer. Several attempts to aerate the egg mixture to prevent densification during processing turned out to be futile, as any air incorporated in the mix was subsequently expelled during HP treatment.

Potato Wedges

Raw red skin potatoes (with skin on) were cut into 1–1.5″ wedges. The wedges were briefly pan fried for about 8 min in vegetable shortening to develop color and flavor. After frying, the potato wedges were sprinkled with salt and a coarse-ground mixed pepper blend.

Green Beans
Fresh raw green beans were trimmed on both ends and packed in brine solution consisting of 1% sodium bicarbonate and 0.9% calcium lactate to provide better flavor, color, and texture. The ratio of brine : green beans was 15 mL per 50 g green beans.

Product Packaging

The chicken and salmon products were packed in 100 g portions, and the potato wedges, eggs, and green beans were packed in 50 g portions. All of the products, with the exception of the egg product, were vacuum packed in clear retort pouches with a SiOx barrier layer (Alcan, Chicago, Illinois). Due to the entrapped gas (from beating) and its foaming tendency, egg was not vacuum packed.

Thermal (Retort) Treatment

Retort processing of all samples was conducted in a horizontal pouch retort (Figure 9.1, Reid Boiler Works Company) located at the US Army Natick Soldier Research, Development, & Engineering Center, Natick, Massachusetts. The capacity of the retort was 37 ft^3. The retort was heated with steam and hot water to achieve the appropriate temperature-time profile needed to ensure the microbiological safety of the product. Preliminary experiments were carried out to determine the time and temperature required for all products to achieve an F_0 of 6 min (Table 9.3).

High Pressure Treatment

All of the high pressure treatments were conducted on a 35-L Avure Technologies HP Unit (Figures 9.2A and 9.2B) located at the National Center for Food Safety & Technology (Summit-Argo, Illinois). This machine has a maximum operating pressure of 690 MPa and a maximum operating temperature of 120°C.

Prior to high pressure treatments, samples were pre-heated to a controlled initial temperature using a two-step procedure. For salmon and egg products, the samples were first equilibrated in a 40°C water bath for 15 min, and then they were pre-heated in a 90°C hot water bath until they reached a temperature of 70°C. For chicken, potato, and green bean

Table 9.3. Processing time and retort temperature for chicken, salmon, potato, green beans, and egg

Product	F_0 (min)	Max. Temperature (°F/°C)	Processing Time (min)
Chicken	7.2–8.8	243/117.2	91
Salmon	7.1–7.6	241/116.1	101
Egg	6.7–7.8	244/117.8	78
Potato	6.9–9.1	242/116.7	93
Green Beans	6.4–9.3	244/117.8	78

Figure 9.1. A horizontal pouch retort (Reid Boiler Works Company, Bellingham, Washington) located at the US Army Natick Soldier Research, Development, & Engineering Center, Natick, Massachusetts.

products, the samples were first equilibrated in a 40°C water bath for 15 min, and then they were pre-heated in a 90°C hot water bath until they reached a temperature of 80°C. Pre-heating the samples in this two-step manner provided the benefits of accelerating the pre-heating process,

Figure 9.2A. 35-L Flow Avure Technologies HP Unit located at the National Center for Food Safety & Technology, Summit-Argo, Illinois.

without inducing protein denaturation and excessive overcooking of the exterior surfaces of the products.

Two different terminal pre-heating temperatures were selected for the two different product groups based primarily on differences in their composition (i.e., fat content of the product) that resulted in attaining similar final temperatures for both groups, when they were subjected to high pressure and concomitant compression heating. Specifically, salmon and eggs have higher fat content than chicken, potato, and green beans. According to published studies (Rasanayagam et. al., 2003), during

Figure 9.2B. 35-L Flow Avure Technologies HP Unit located at the National Center for Food Safety & Technology (Summit-Argo, Illinois).

adiabatic heating, fats experience a higher temperature rise than water under compression. Therefore, due to compression heating at 690 MPa, salmon and eggs will experience a greater temperature rise than potato, green beans, and chicken. The lower pre-heating conditions for the salmon and egg (70°C) compared to chicken, potato wedges, and green beans (80°C) tended to minimize these differences so that all products would experience similar conditions during processing.

The high pressure treatments practiced in these experiments were done according to the two-pulse process taught by Meyer (US Patent 6,177,115). In the two-pulse process, the samples were processed to a maximum pressure of 690 MPa and held at pressure for 1 min for each of two consecutive pressure pulse sequences. The pressure was released and returned to ambient conditions (1 atmosphere) between pulses. The come-up time to reach 690 MPa was approximately 90 seconds for each HP pulse. The peak temperature attained at pressure for all of the products was approximately 106 °C.

After processing with HP or retorting, all of the samples were stored in refrigeration (4°C) to ensure food safety (or prevent growth, in the event that full sterility was not achieved) and to minimize changes in properties of the products after processing. Analyses were conducted within 2 weeks of preparation to ascertain physical and sensory characteristics of the products that would most likely distinguish the effects of HP versus retort processing.

Quality Evaluations

Sensory Measurements

Important quality attributes such as flavor, color, and texture were evaluated in these sensory experiments to determine the differences between HP sterilized and retorted food products that could be perceived by potential consumers. Trained panelists were selected to conduct the sensory evaluations. Paired comparison/descriptive evaluations were used to assess the quality attributes of appearance, flavor, taste, and texture of the food products. A pair of coded samples (HP and retort) from each food category was presented to the trained panelists for comparison. Panelists were asked to describe the quality attributes and register differences between the samples.

Color Measurements

A Minolta Color Meter (Minolta Spectrophotometer CM-2002, Minolta Camera Co., Japan) was used to determine the L* (lightness), a* (redness or greenness), and b* (yellowness or blueness) values of each food product. Prior to the L*, a*, b* measurements, standard white and black tiles were used to calibrate the color meter. In order to obtain a good overall representation of the L*, a*, b* values, ten readings from ten different locations of each sample were obtained. After each set of ten readings, the standard white and black tiles were measured again to ensure there was no drift in the instrumental calibration.

Texture Measurements

Texture measurements on each sample after HP sterilization or retorting were made using shear tests. Prior to the texture measurements, salmon, egg, chicken, and potato samples were cut into cubes (1.5 cm × 1.5 cm × 1.5 cm). Green beans were cut into 5 cm lengths. The texture of each sample was evaluated using a single blade shear cell (10 cm × 5 cm × 0.3 cm) on a TA.XT2 Texture Analyzer (Texture Technologies Corp., Scarsdale, New York/Stable Micro Systems, Godalming, Surrey, UK). Shear tests were performed at a cross-head speed of 3 mm/s. Three replicates of the texture measurements were obtained for each sample as the recorded plots of shear force (g) versus distance (mm).

Results and Discussion

Chicken

Sensory Evaluations

According to the panelists, the retorted chicken had a flavor similar to stewed or canned chicken, whereas the HP-treated chicken retained the flavor of a fresh roast chicken. The texture of the retorted chicken was described as fibrous, tender, and soft, and, in contradistinction, the HP-treated chicken was described as being moist and tender. Figure 9.3A shows that there was a distinctive difference observable in the muscle structure of retorted versus HP-treated chicken. The retorted chicken showed a loss of muscle integrity, whereas the HP-treated chicken featured a still-intact muscle structure.

Color Evaluations

The L*, a*, b* values indicated that HP chicken samples were lighter and more yellow in color than thermally treated chicken (Figure 9.3B).

Texture Studies

The TA.XT2 single blade shear measurements showed that both the HP and thermally treated chicken had similar resistance to cutting done

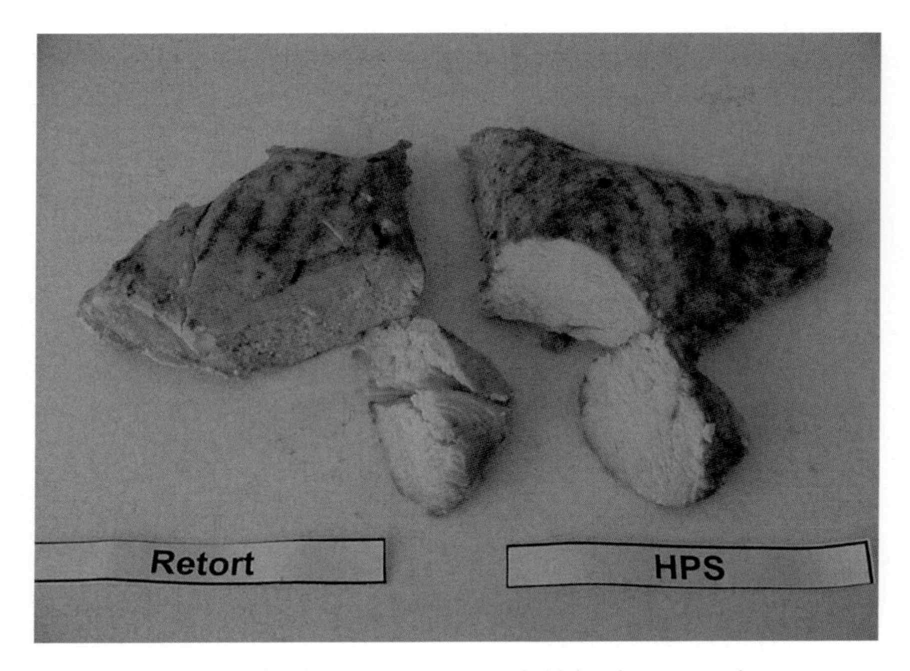

Figure 9.3A. Retorted and high pressure treated chicken breast samples.

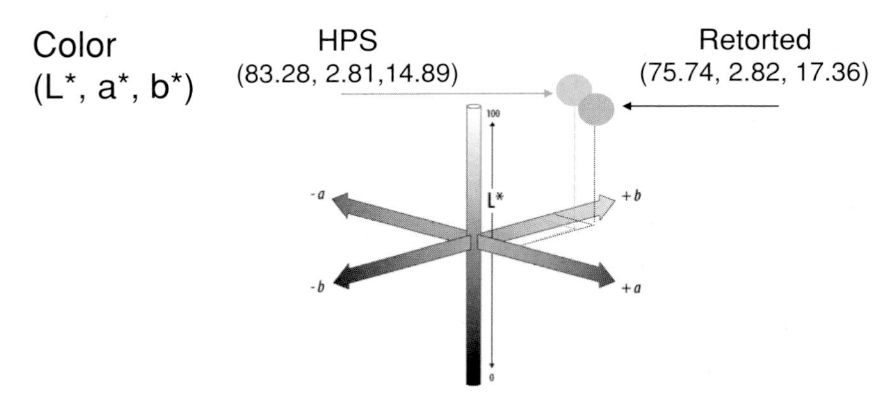

Figure 9.3B. L*, a*, b* values of high pressure sterilized (HPS) versus retorted chicken breast samples. The HP-treated chicken was lighter and more yellow than its retorted counterpart.

perpendicular to the grain of the muscle fibers (Figure 9.3C). Both samples were thus similarly tender. However, a loss of meat fiber cohesiveness was evident in the retorted chicken. During the texture measurements, the retorted chicken tended to "fall apart" parallel to the

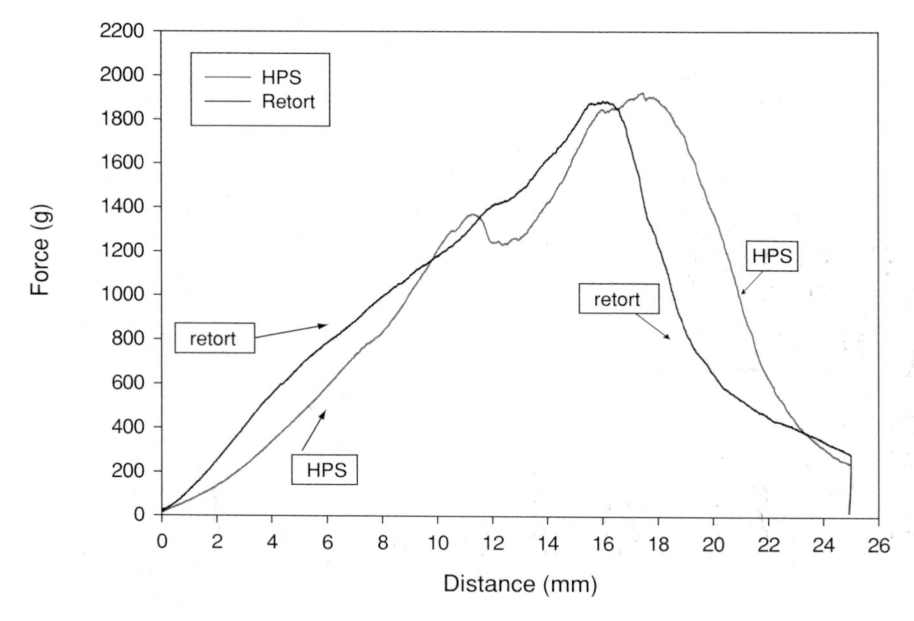

Figure 9.3C. Plot from single blade shear measurements of HP-sterilized (HPS, red line) and retorted (blue line) chicken.

grain and in sensory evaluations provided what the panelists termed less "chew".

Salmon

Sensory Evaluation

The appearance of the retorted salmon was unappealing to our taste panelists (Figure 9.4A). The panelists associated the faded pink color of the retorted salmon with overcooking, and the flesh was very soft and showed evidence of oil exudation. The tissue of the retorted salmon started to fall apart upon cutting, and there was generally a loss of the "flaky" texture normally attributed to properly cooked fish.

The taste panelists indicated that the HP-treated salmon samples, on the other hand, featured a bright pink color, were not oily, and exhibited characteristics commonly associated with freshly poached salmon: moistness, flaky texture, and having good flesh integrity.

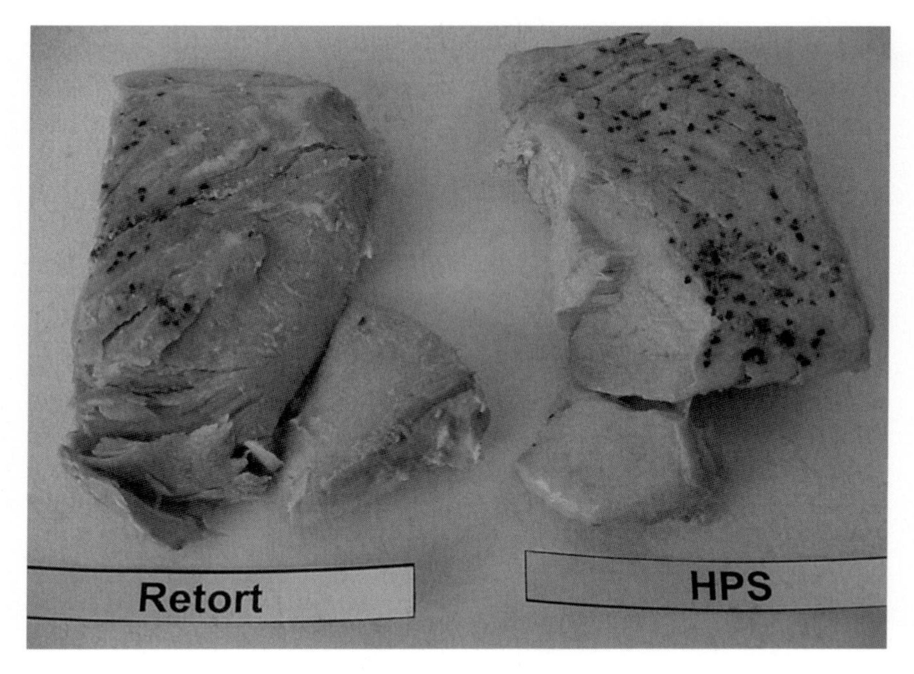

Figure 9.4A. Retorted and high pressure treated salmon.

Color Evaluations

The L*, a*, b* values indicated that HP-treated salmon was lighter and more orange than the retorted salmon (Figure 9.4B).

Texture Studies

The TA.XT2 single blade shear measurements showed that HP-treated salmon samples had a greater resistance to cutting perpendicularly to the grain, and thus it was firmer than the retorted salmon (Figure 9.4C). The loss of structure in the fish tissue was evident in the retorted salmon. When the blade cut through the retorted salmon, the flesh fell apart immediately. Such soft texture is associated with retorted salmon.

Potato Wedges

Sensory Evaluation

The retorted potatoes had mild musty and overcooked notes. Some browning and discoloration of the surface was observed with the retorted

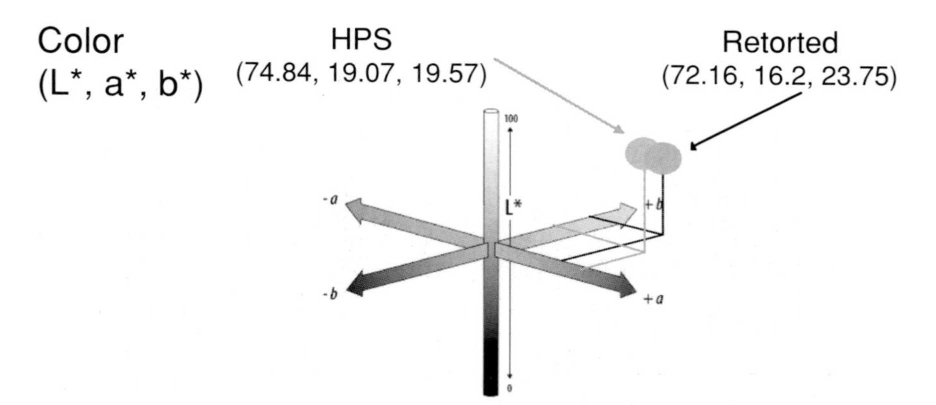

Figure 9.4B. L*, a*, b* values for retorted and high pressure treated salmon.

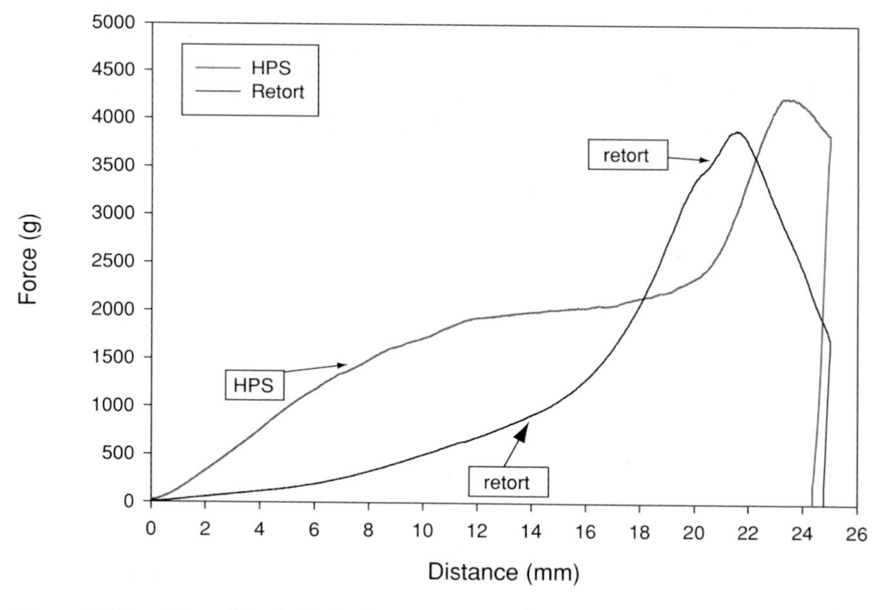

Figure 9.4C. Plot of single blade shear measurements from retorted and high pressure treated salmon.

potatoes that could not be attributed to the pre-frying (Figure 9.5A). The retorted potatoes had a firm crumbly texture and developed a dry skin on the cut surfaces.

The HP-treated potatoes had a fresh boiled-like flavor, a smooth, tender, and mealy texture, and also some fried and spicy flavor notes. The surface of the potatoes was whiter and was not discolored.

Figure 9.5A. Retorted and high pressure treated potato wedges.

Color Evaluation
The L*, a*, b* values indicated that retorted potato wedges were darker and more brown than the HP-treated samples (Figure 9.5B).

Texture Studies
The TA.XT2 single blade shear measurements showed that both samples had similar peak resistance to cutting. The retorted samples had a firm outer surface and fractured rapidly when the blade touched the samples (Figure 9.5C). HP-treated potatoes showed less surface firmness and allowed greater penetration before fracturing than the retorted potatoes. The retorted potato wedges exhibited a thin, dry leathery skin on the surface.

Egg

Sensory Evaluations
The taste panelists first noted that the retorted egg products had a dull yellow color and also a greenish-brown hue. This greening of the

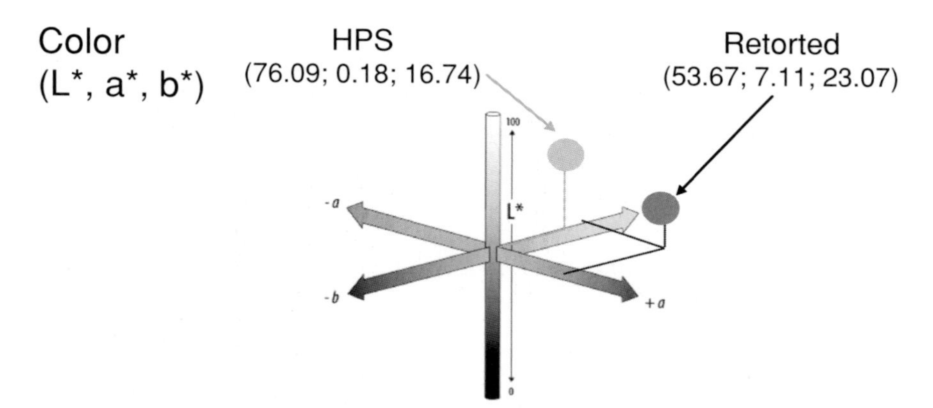

Figure 9.5B. L*, a*, b* values for retorted and high pressure treated potato wedges.

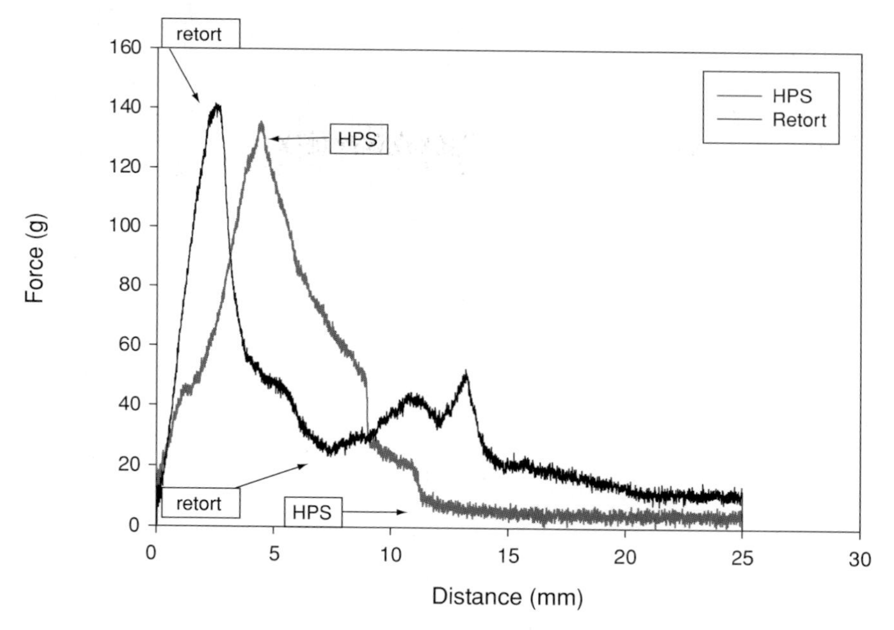

Figure 9.5C. Single blade shear measurements for retorted and high pressure treated potato wedges.

retorted egg products is usually associated with excessive thermal processing. The greening of the present egg products still occurred for the retorted eggs, irrespective of the addition of citric acid to the egg products. Some metallic and sulfury notes were also detected when sampled by the taste panels. The texture of the retorted eggs was soft and somewhat aerated (the presence of air bubbles is visible in Figure 9.6A).

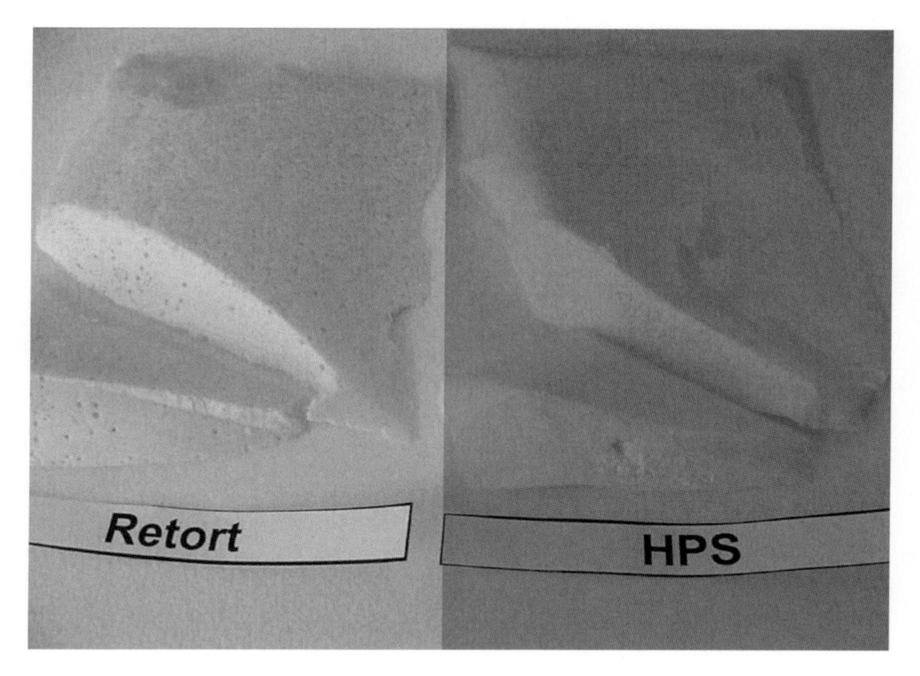

Figure 9.6A. Retorted and high pressure treated egg samples.

The HP-treated egg products appeared to have good color, a fresh hard-boiled egg flavor, and a custard-like firm texture. The texture of the HP-treated egg product was creamy and dense, without the presence of air bubbles.

Color Evaluations
The L*, a*, b* values indicated that the retorted egg was darker and more yellow/green than the HP-treated eggs (Figure 9.6B).

Texture Studies
The TA.XT2 single blade shear measurements showed that the HP-treated egg products had a greater resistance to cutting and also a very firm texture (Figure 9.6C). The retorted eggs cut easily and had a softer texture. In addition, as mentioned above, the retorted eggs also exhibited porosity (gas bubbles that contributed to softening the texture). Gas bubbles were not present in the HP-treated eggs, which contributed to its denser, custard-like texture.

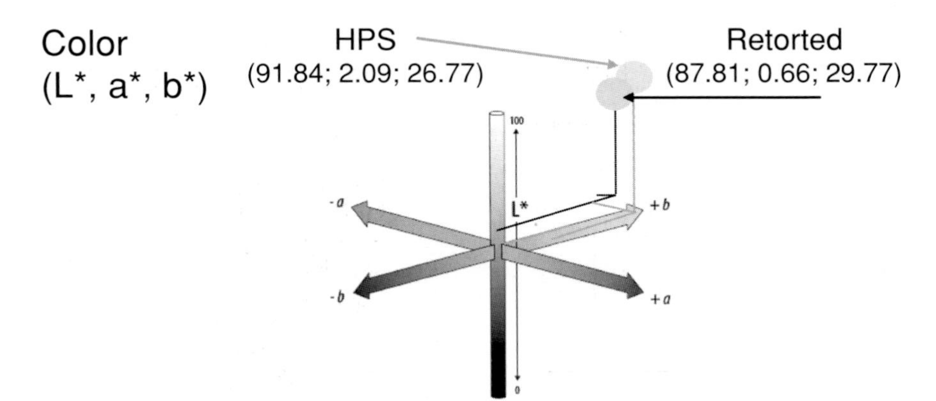

Figure 9.6B. L*, a*, b* values for retorted and high pressure treated egg samples.

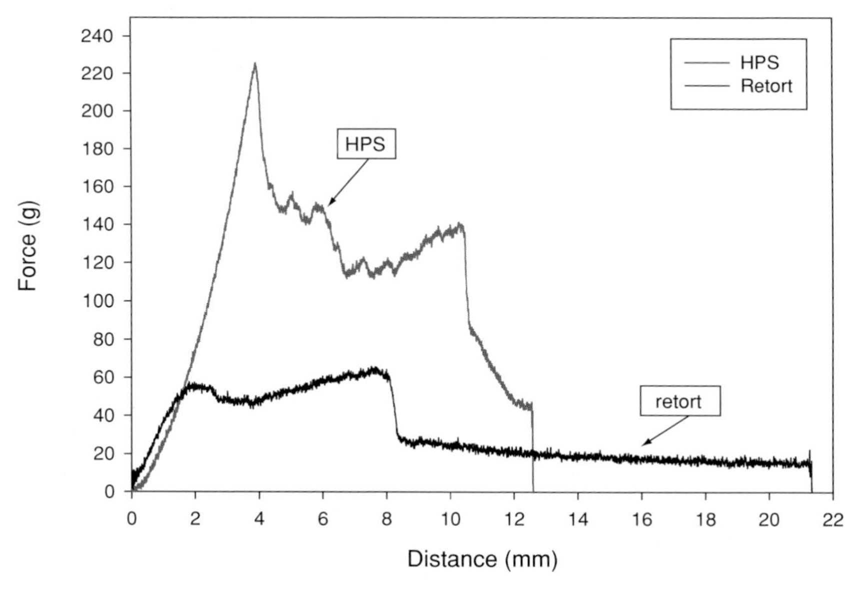

Figure 9.6C. Single blade shear measurements for retorted and high pressure treated egg samples.

Green Beans

Sensory Evaluations

The retorted green beans had the taste and texture typically exhibited by commercial canned green beans. The retorted green beans tended to

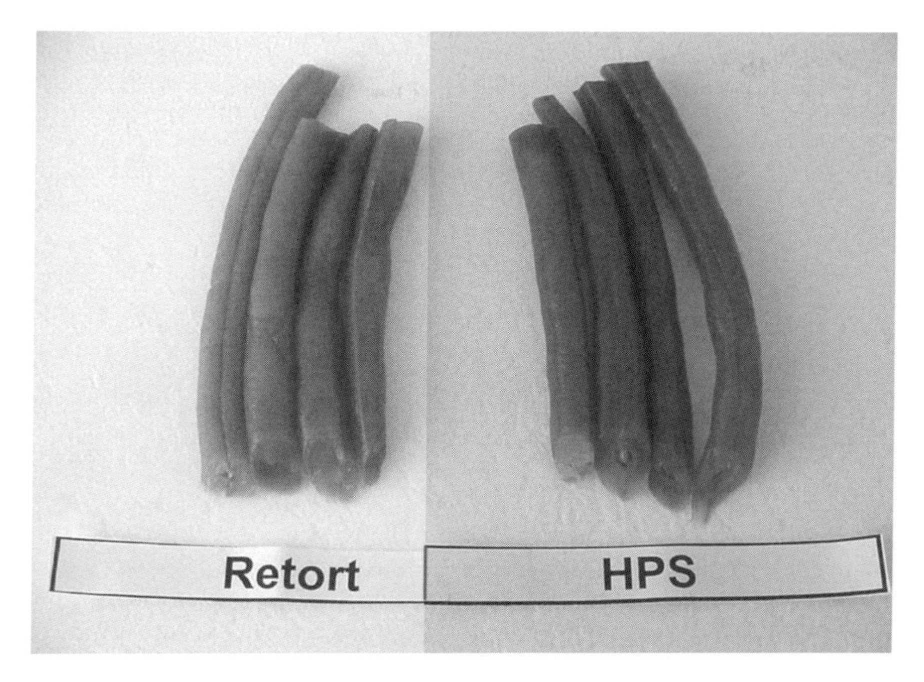

Figure 9.7A. Retorted and high pressure treated green beans.

separate at their seams upon handling, indicating that disintegration of the tissue had occurred during heat treatment (see sliced end of beans in Figure 9.7A). The taste panels described the texture of the retorted green beans as very soft, and the green beans did not require much chewing. The taste panels also observed that the HP-treated green beans had a discernible darker color than the retorted green beans. Their texture was very crunchy, suggesting a firmness similar to that of lightly steamed green beans, and the sliced ends reveal the retention of inner structure. The HPS beans also exhibited a slightly raw and earthy flavor.

Color Evaluations

The L*, a*, b* values indicated that the HP-treated green beans were darker than the retorted green beans (Figure 9.7A). The color measurements made with the Minolta were made on several beans laid side to side to cover the aperture. Numbers may be inaccurate due to influence of dark spaces in regions between the beans (Figure 9.7B).

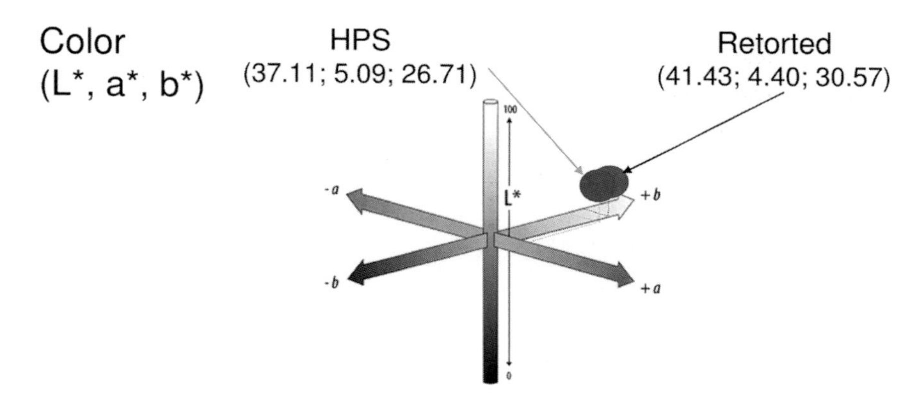

Figure 9.7B. L*, a*, b* values for retorted and high pressure treated green beans.

Texture Studies

The TA.XT2 single blade shear measurements showed that HP-treated green beans had a significantly greater resistance to cutting and a very firm texture. In contrast, the retorted green beans cut easily and were very soft. The retorted green beans were visibly swollen, soft, and easily separable at the seams if handled (Figure 9.7C).

Discussion and Conclusions

The high pressure sterilization process employed in the present studies has yet to be validated as adequate for achieving food safety and shelf stability. The samples prepared in this study and held at room temperature showed no evidence of spoilage over a 3-month storage period. The raw material selection, pre-treatment methods, pre-heating conditions (temperatures and times), and HP processing conditions (high pressure, temperature, hold times, number of high pressure pulses) are all expected to influence the quality attributes of the finished products. However, based on this limited study using the processing conditions described above (that are believed, but not yet proven or validated), we cautiously draw these conclusions for this study:

- HPS delivers fresher, less processed flavor in all of the foodstuffs tested as a result of less total thermal exposure than traditional retorting.

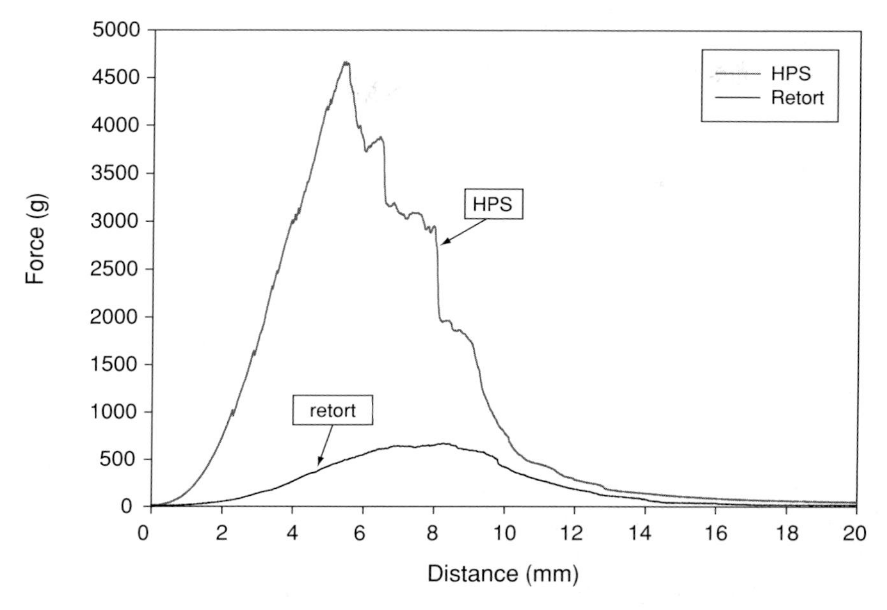

Figure 9.7C. Single blade shear measurements for retorted and high pressure treated green beans.

- HPS results in less caramelization/browning, and, in most cases, leads to lighter colored products (with the exception being green beans, which do not retain the color of green beans under the conditions employed in this study).
- The effect of HPS on texture varies and depends on the foodstuff:
  - Pre-testing indicates a tendency toward more firmness in HPS treatments of protein-based foods and vegetables.
  - Whole muscle structure is better retained with HPS, but it can lead to rubberiness in some foods.
  - Retorting achieves softness but at the expense of texture. Connective tissue between fibers is solubilized to a greater extent in retorting than in HPS.
  - The firmer texture of HPS egg is attributed primarily to the inability to retain a foamy structure through processing.
- HPS shows promise as a technology that can deliver products of significantly better quality than retorting:
  - These results indicate that in general HPS foods have better sensory attributes than retorted.

- Our pre-testing indicated that the texture of HPS foods can be optimized by raw material selection and pre-treatment, allowing the full quality potential of HPS to be realized.
- Pre-testing also indicated that the pre-heating process has a large influence on final product quality. To the extent that pre-heating can be accelerated, quality will benefit.
- Higher pressure operation (and lower initial product temperatures) would also enhance the quality of foods made with HPS.

Acknowledgments

The authors would like to express their thanks to the following persons for their assistance in various aspects of this research study: Dr. Richard Meyer (Washington Farms); Dr. Tom C.S. Yang, Dr. C. Patrick Dunne, and Mr. Jay Jones (US Army Natick Soldier Research, Development, & Engineering Center); Mr. Eduardo Patazca (National Center for Food Safety & Technology); and Dr. Mahesh Padmanabhan, Dr. Andrew McPherson, and Mr. Leopold Young (Kraft Foods).

References

Galazka, V.B., and D.A. Ledward. 1995. Developments in high pressure food processing. *Food Technology International Europe* 12:123–125.

Hayashi, R. 1989. "Application of high pressure to food processing and preservation: Philosophy and development." In: *Engineering and Food,* ed. W.E.L. Spiess and H. Schubert, pp. 815–826. London: Elsevier Applied Science.

Kimura, K, M.Ida, Y. Yoshida, K. Ohki, T. Fukumoto, and N. Sakui. 1994. Comparison of keeping quality between pressure-processed and heat processed jam: Changes in flavor components, hue and nutritional elements during storage. *Bioscience Biotechnology Biochemistry* 58:1386–1391.

Knorr, D. 1993. Effects of high-hydrostatic-pressure processes on food safety and quality. *Food Technology* 47:156–161.

Mertens, B. 1992. Under pressure. *Food Manufacture* 11:23–24.

Meyer, R.S. 2000. Ultra high pressure, high temperature food preservation process. US Patent 6,017,572.

———. 2001. Ultra high pressure, high temperature food preservation process. US Patent 6,177,115 B1.

Parish, M.E. 1998. Orange juice quality after treatment by thermal pasteurization and isostatic high pressure. *Lebensm. Wiss. Technol.* 31:439–442.

Rasanayagam, V., V.M. Balasubramaniam, E. Ting, C.E. Sizer, C. Anderson, and C. Bush. 2003. Compression heating of selected fatty food substances during high pressure processing. *Journal of Food Science* 68(1):254–259.

Thakur, B.R., and P.E. Nelson. 1998. High pressure processing and preservation of food. *Food Review International* 14:427–447.

Wilson, M.J., and R. Baker. 2000. High temperature/ultra-high pressure sterilization of foods. US Patent 6,086,936.

———. 2001. High temperature/ultra high pressure sterilization of foods. US Patent 6,207,215.

Yen, G.C., and H.T. Lin. 1996. Comparison of high pressure treatment and thermal pasteurization effects on the quality and shelf life of guava puree. *International Journal Food Science and Technology* 31:205–213.

———. 1999. Changes in volatile flavor components of guava juice with high pressure treatment and heat processing and during storage. *Journal of Agriculture Food Chemistry* 47:2082–2087.

Chapter 10

Consumer Evaluations of High Pressure Processed Foods

Alan O. Wright, Armand V. Cardello, and Rick Bell

Introduction

The US Army Natick Soldier Research, Development, & Engineering Center (NSRDEC) has played a central role in the development of novel food processing technologies for the preservation of foodstuffs intended for use in military rations and for commercialization. There is a continuous effort to find new methods for producing safe, nutritious, appealing rations that either feature improvements in quality and consumer acceptance compared to conventionally processed foods, or expand the selection and variety of menu items available in military rations. Nonthermal technologies provide a rich array of food processing methods to produce safe, higher quality, value-added foods that feature higher vitamin retention, improved sensory attributes (appearance, flavor, aroma, texture, etc.), and higher consumer acceptance than their thermally processed counterparts. In this chapter we explore consumer attitudes and perceptions toward foods treated with the emerging technology of high pressure processing (HPP) that establish a basis to guide food technologists in the development and commercialization of new products.

Consumer Concerns of High Pressure Processing

In a study in which 25% of some consumer groups did not understand the meaning of the term "high pressure processing," Deliza et al. (2003)

Table 10.1. Percent of (n = 198) respondents that were "very" or "extremely" concerned with foods processed by novel techniques

Food Processing Method	% Very or Extremely Concerned	% Uncertain
Genetically modified	54	17
Irradiation	49	17
Radio frequency sterilization	40	21
High pressure treatment	20	18
Microwave processing	18	12
Thermal processing	18	14
Heat pasteurization	13	6

showed that this lack of familiarity translated into a negative perception of acceptance of HPP-treated pineapple juice and a lower purchase intent. In a separate study as part of a broad research program at NSRDEC to assess consumer attitudes toward novel, nonthermal processing techniques, Cardello (2000) surveyed 198 military consumers for their level of concern with eating foods treated by different food processing technologies (Table 10.1).

Genetic modification and irradiation of foods evoked the most concern among respondents (54% and 49% of respondents, respectively, were either "very" or "extremely" concerned with foods preserved in these manners). Conventional thermal processing (18%) and heat pasteurization (13%) generated significantly less concern. HPP was rated at 20%, which is much closer to the level of concern shown for thermal processing than that of genetic modification or irradiation. As the respondents were consumers, it can be assumed that they were mostly unfamiliar with the technical aspects of each of the individual technologies. However, the relatively low level of concern expressed by them toward the label "high pressure treatment" implies that food industry leaders would most likely not encounter consumer resistance in the marketing of HPP-treated foods. By comparison, the strong adverse reaction of the public toward irradiated or genetically modified food labels has been the object of much discussion in the media.

In a preliminary study conducted by NSRDEC researchers, two groups of respondents differing in their familiarity with HPP technologies evaluated the same set of HPP-treated products. One group

Table 10.2. Value-added benefits of different processing methods

	Labeled Affective Magnitude +100 (most favorable) to −100
Nutritional health benefit	
HPP	80.26
Microwave	47.97
Thermal	33.29
Quality benefit	
HPP	74.45
Microwave	45.33
Thermal	40.8
Purchase price benefit	
Thermal	51.84
Microwave	47.07
HPP	44.9

consisted of "motivated consumers" (international representatives of food companies with an active interest in the development of HPP), and the other group consisted of average consumers in the United States. In general, the U.S. consumers were less familiar with HPP and typically rated the products lower than did the "motivated consumers." These "motivated consumers" represent an exceptional subpopulace that includes individuals with a vested interest in the marketing of HPP foods. These "motivated consumers" expressed considerable optimism regarding the potential for commercialization of HPP products in the next 7 years based on gains in nutritional content and quality that will offset any slight increases in price (Table 10.2). It is important to remember that typical consumers comprise the majority of the marketplace, and these consumers ultimately will determine the success of new products, in large part based on whether they perceive HPP to impart products with some noteworthy advantages (Butz et al., 2003).

Providing Information to Shape Consumer Expectations

Providing consumers with prior information of a product can help "frame" their responses to actually receiving the product (Kahneman and Tversky, 1974; Tversky and Kahneman, 1984). The information presented to the consumers is intended to help consumers form expectations of that product (Cardello, 1994, 2007). These expectations can influence consumers' perceptions and liking of the product, and it is these expectations that are subsequently either confirmed or disconfirmed by the actual product. Predictive models relating the influence of expectations on consumer perceptions and the acceptance of foods have been reported (summarized by Cardello, 1994 and 2007, and by Deliza and MacFie, 1996).

A study recently conducted at NSRDEC (Cardello, 2003) examined the relationship of consumers' liking of foods processed by different technologies in the context of their concerns and "framed" expectations. Volunteer consumers rated thermally processed chocolate pudding products in different conditions of provided information. First, subjects rated their potential liking of the chocolate pudding (no samples or other information provided) and their level of concern for different processing technologies (e.g., irradiation, high voltage pulses, high pressure, pulsed electric field, nonthermal preservation, and the addition of bacteriocins). In a second session conducted several weeks later, consumers were provided different sets of information along with pudding samples designated for sensory evaluation. The information consisted of either (i) being told the pudding had been processed by a novel food processing technique (without specifically naming the technique), (ii) the name of the novel technology and an objective description of that technology, or (iii) the name of the technology, an objective description, and a statement of the benefit. In all cases, the subjects rated the products three times: (i) after receiving the information, (ii) after receiving the information and observing the sample, and (iii) after receiving the information, observing the sample, and tasting the sample.

For all three information conditions, the "concern" ratings for the different food processing technologies had negative correlations with expectations of liking for the chocolate pudding that consumers *believed* was processed by an alternative technology. In addition, the lowered expectations resulted in decreased liking of the pudding. These results support the premise that consumers' perceptions about the risk of a technology significantly influenced their actual liking of food products processed by that technology.

The study also demonstrated that factual information about the product and process, product exposure (seeing the product), and a simple safety/benefit statement all contributed to increasing the consumer's expected liking and acceptance of the product. For HPP, allowing consumers to see the product increased their acceptance of the product by 0.37 points on a 9-point scale. Similarly, providing consumers a description of the technology increased their expected liking of the product by 0.08 scale points, and providing consumers a benefit statement of the technology added an average of 0.23 scale points. Since a difference of 0.5 scale points will typically produce a significant difference in consumer acceptance ratings between products on a 9-point scale, these simple strategies of informing and familiarizing consumers (which added a combined total of 0.68 scale points) have the potential to significantly increase the expected and measurable liking of nonthermally processed foods.

Consumer Perceptions of High Pressure Processing

In a recent set of studies conducted at NSRDEC, Cardello et al. (2007) used conjoint analysis to assess the importance of various product and marketing factors on consumer interest in foods processed by emerging technologies for three different user groups. Conjoint analysis is a sophisticated survey technique in which consumers evaluate multiple product concepts that differ in variables of interest to the product developer. The results of a conjoint analysis allow product developers to assess consumer interest in the use of a product. Figure 10.1 (Cardello et al., 2007) shows the relative contribution of seven different food processing techniques to consumer interest in the use of them. On the ordinate, positive "utility" values indicate greater interest in potential use of the product by the consumer, while negative values indicate lower interest. These data show that HPP has a strong positive influence on consumer interest, while, for comparative purposes, irradiation and genetic modification have extremely low interest among consumers.

Consumer Perceptions and Sensory Analysis of High Pressure Processing

While HPP has the potential for producing value-added food products, it may not, like many other processing technologies, be the best

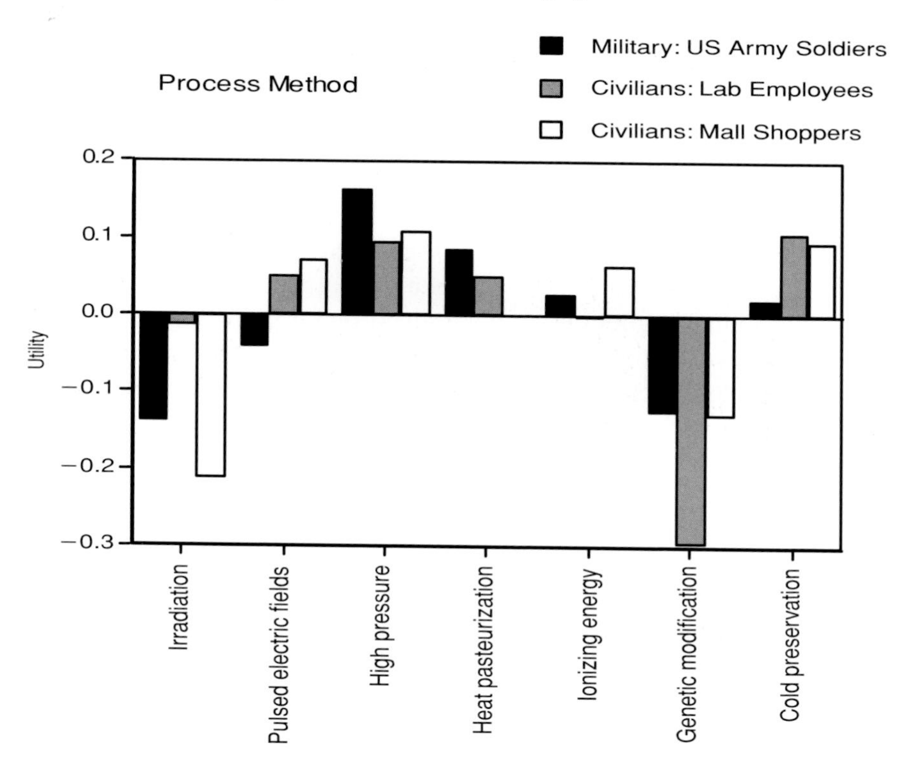

Figure 10.1. Processing method utility graph (Cardello et al., 2007).

alternative for all food products, food types, or consumer market sectors. For example, egg products are a difficult commodity to deal with in terms of receiving and retaining high ratings over prolonged storage conditions (3 years at 80°F to meet the Army's basic field ration requirement). In fact, fresh scrambled eggs typically score around a 7 on a 9-point quality scale (attribute trained panelist) or a 9-point Hedonic scale (untrained, acceptability rated). The conventional method of thermal processing to make shelf-stable egg products is often rated (5-rating) 2 full points lower on a 9-point scale than fresh scrambled eggs. It is important to note that, on either the quality or Hedonic scale, a 5.5 or less is generally considered unacceptable for military rations. Accordingly, egg products appeared to be an ideal candidate for evaluation with a novel technology such as HPP, and NSRDEC researchers explored the potential for developing improved egg products using HPP.

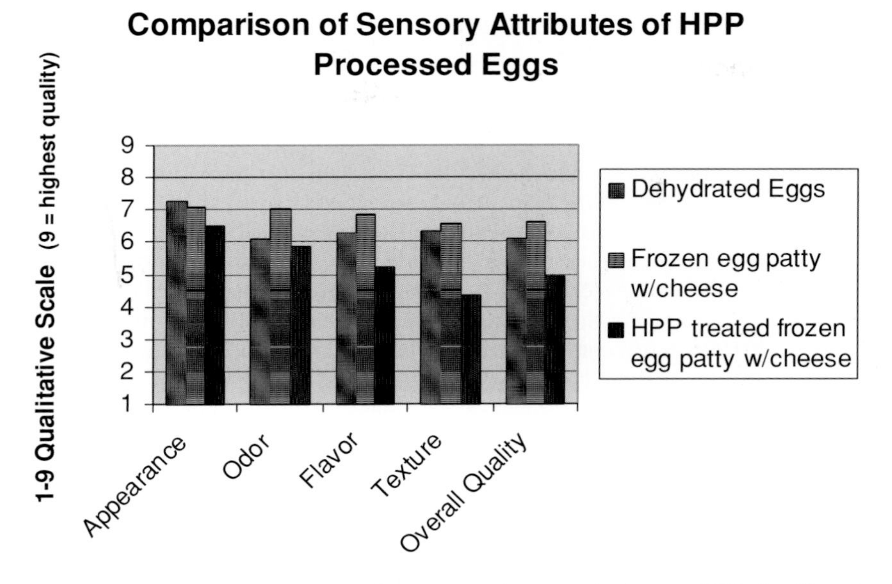

Figure 10.2. Sensory evaluation of HPP-treated, pre-cooked eggs. (HPP significantly different from all samples and attributes ($P = 0.0001$.)

Figure 10.2 shows the sensory ratings in terms of appearance, odor, flavor, texture, and overall quality for three egg products: dehydrated eggs (a military ration item), a commercially available pre-cooked (stored frozen and re-heated prior to serving) egg and cheese patty, and the HPP-treated egg and cheese patty in conditions to render a shelf-stable (i.e., commercially sterile) product ($P = 700$ MPa, $T = 105°C$, $t = 5$ min).

Dehydrated eggs received a rating around ~6.0 for most categories, and ~7 for appearance. The commercially available pre-cooked, re-heated egg and cheese patty received consistently high ratings in the range of 6.5–7.0. Treating these commercial egg and cheese patties with HPP to make a shelf-stable product resulted in decreased acceptance of the product below a minimally acceptable level (overall quality ~5.0). The texture (~4.3) was a particular detractor for the HPP product. As noted in a previous chapter, HPP treatments of eggs collapse the porous, air-filled voids and cell structure of the fluffy eggs with a resultant densified texture. Despite these obvious challenges, more research work is planned for this area, to eventually expand the benefits of HPP for eggs and other breakfast foods.

References

Butz, P., E.C. Needs, A. Baron, O. Bayer, B. Geisel, B. Gupta, U. Oltersdorf, and B. Tauscher. 2003. Consumer attitudes to high pressure processing. *Food, Agriculture and Environment.* 1(1):30–34.

Cardello, A.V. 1994. Consumer expectations and their role in food acceptance. In: *Measurement of Food Preferences,* ed. H.J.H. MacFie and D.M.H. Thomson, pp. 253–297. Glasgow: Blackie Academic and Professional.

———. 2000. Consumer attitudes and expectations toward non-thermal and other novel food processing techniques. Presentation at the IFT Non-thermal Processing Division Workshop on Non-thermal Processing of Food. Portland, OR.

———. 2003. Consumer concerns and expectations about novel food processing technologies: Effects on product liking. *Appetite* 40:217–233.

———. 2007. "Measuring consumer expectations to improve product development." In: *Consumer-Led Food Product Development*, ed. H.J.H MacFie, pp. 223–261. Cambridge, UK: Woodhead.

Cardello, A.V., H.G. Schutz, and L.L. Lesher. 2007. Consumer perceptions of foods processed by innovative and emerging technologies: A conjoint analytic study. *Innovative Food Science and Emerging Technology* 8(1):73–83.

Deliza, R., and H.J.H. MacFie. 1996. The generation of sensory expectation by external cues and its effect on sensory perception and hedonic ratings: A review. *J. Sensory Studies* 11(2):103–128.

Deliza, R., A. Rosenthal, and A.L.S. Silva. 2003. Consumer attitudes towards information on non-conventional technology. *Trends in Food Science and Technology* 14:43–49.

Kahneman, D., and A. Tversky. 1974. *Judgment Under Uncertainty: Heuristics and Biases.* Cambridge, MA: Cambridge University Press.

Tversky, A., and D. Kahneman. 1984. Choices, values and frames. *American Psychologist* 39:341–350.

Chapter 11

Compression Heating and Temperature Control in High Pressure Processing

Edmund Ting

Introduction

The study of microbiological inactivation under high hydrostatic pressure for the safety and preservation of foods requires the generation of data under accurate pressure and temperature conditions. While pressure is easily controlled, constant temperature conditions are difficult to attain, primarily due to the phenomenon called compression heating. The primary error associated with the lack of temperature control is a non-constant lethality rate that results in artificial microbiological tailing. While microbial inactivation kinetics are known to exhibit tailing effects under some conditions of constant lethality, the occurrence of artificial tailing can confound the analysis of the inactivation kinetics observed with high pressure processing (HPP).

Rapid pressurization of compressible substances results in an unavoidable, thermodynamically induced temperature change. The maximum change in temperature occurs in adiabatic conditions. The more compressible a substance, the greater is its level of adiabatic heating. Most foods exhibit compression heating characteristics similar to those of water, with an adiabatic heating value of 3–4°C/100 MPa, depending on the initial temperature (Rasanayagam et al., 2003). At higher initial temperatures, water is more compressible, which results in an even greater extent of compression heating than that which occurs at lower temperatures and the same level of pressurization. During rapid compression, the compressible contents of the vessel (foods,

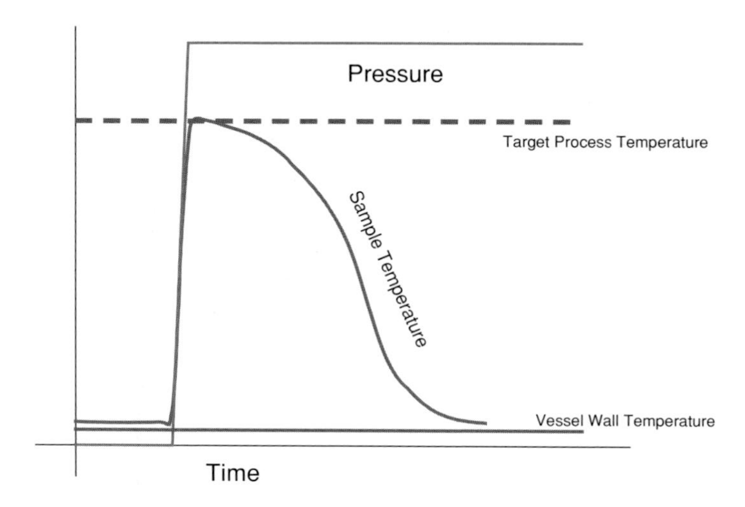

Figure 11.1. Isothermal start approach results in rapid temperature drop after pressure come-up.

microorganisms, and water) increase in temperature, but the incompressible metal pressure vessel does not. Accordingly, the temperature of the vessel contents rises to an elevated level, and heat gradually dissipates to the metal vessel (effectively acting as a heat sink). Heat loss in this manner can compromise the effectiveness of a high pressure process by allowing the sample temperature to drop below the level of lethality necessary to ensure the inactivation of relevant microorganisms and the safety of the product to the consumer. Different approaches that can lessen the temperature changes occurring during test conditions are discussed below, with attention on modulating the effects of compression heating and improving the temperature stability of HPP.

Isothermal Start—Rapid Come-Up

The isothermal start approach (shown in Figure 11.1) is useful for tests with rapid pressure come-up rates and short maximum pressure hold times.

For these short-duration experiments, there is insufficient time for significant heat transfer, and the process remains essentially adiabatic.

The effect of sample cooling (transferring heat to the vessel) becomes inevitable for experimental conditions exceeding a few seconds. This effect is commonly observed with the smaller pressure vessels used for microbiological studies that have a relatively large ratio of internal vessel surface area to vessel volume. This limitation becomes less significant as the diameter of the equipment increases. Large commercial processing vessels typically have diameters greater than 7″, and the heat transfer losses become less significant, particularly under the test conditions of interest (processing times < 3 minutes). Insulating materials can also reduce heat transfer from the sample to the vessel.

Small Equipment Approaches

Although larger high pressure equipment most closely approximates conditions of constant temperature, it is unfortunately not always available or efficient to use for extensive small-scale testing. Microbiological experiments using smaller equipment should adopt approaches that minimize heat transfer. Recognizing that the cause of heat loss from the sample is attributable to temperature gradients arising from differences in compression heating, the following three methods offer potential solutions to eliminating this difference.

Slow Pressurization while at Process Temperature

In order to decrease the magnitude of compression heating and the resultant differences in the relative temperatures of the sample and vessel wall, a slow pressurization rate is used. Initially, the vessel temperature is maintained at the desired process temperature, and the pressure is gradually increased to its intended final value. The slow pressurization process rate allows sufficient time for compression heating of the sample to dissipate to the vessel without building up a substantial temperature gradient, and the sample and vessel remain close to temperature equilibrium. Unfortunately, the slow compression rates this method requires prolong the come-up time and the overall processing time (Figure 11.2) to scales that are impractical for commercial processes (come-up times and maximum pressure hold times are typically less than 3 min each for these applications).

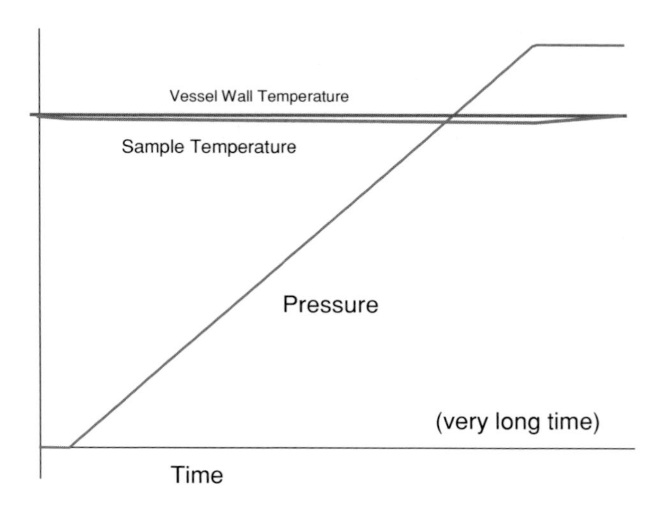

Figure 11.2. Slow pressurization at process temperature is effective for small samples but makes for long experiments.

Changing Vessel Temperature during Pressure

Another approach for creating constant temperatures and pressures is to maintain a constant pressure over the sample and change the temperature of the vessel to match the sample temperature during the come-up time of the pressurization process (Figure 11.3).

Using a pressure source with pressure compensation ability prevents thermal expansion of the content from causing a pressure change. This approach works for systems that use small diameter high pressure tubing as the pressure vessel. For a large sample, the time needed to change the temperature of a high mass pressure vessel is too long for practical interest.

Dynamic Temperature Method (DTM)—Different Initial Temperatures for Sample and Vessel

The DTM involves pressurizing a sample that is held at a significantly lower temperature than the vessel, such that compression heating of the sample induces a rise in sample temperature that is compatible with the vessel temperature (Figure 11.4).

After reaching pressure, the temperature difference between the sample and vessel will be minor and will correspondingly reduce the flow

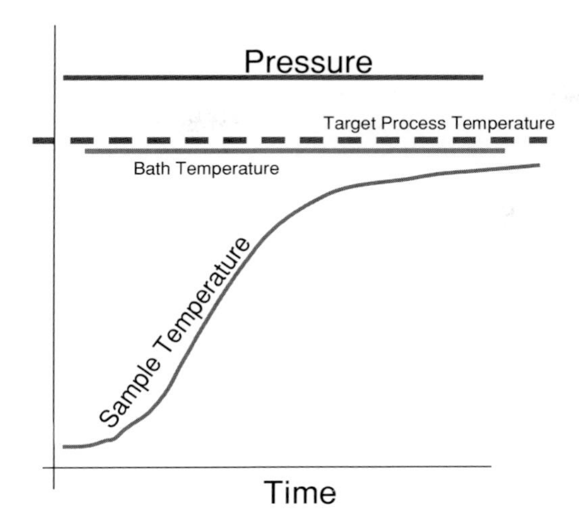

Figure 11.3. Immersion of pressure vessel into temperature bath is feasible only for very small vessels. The thermal mass of most pressure vessels will result in very slow temperature changes.

of heat between the sample and the vessel wall. Determining the appropriate initial temperature (pre-pressurization) is done empirically. A first approximation is estimated by subtracting from the process temperature the expected adiabatic heating change and an estimate of some additional temperature gained by heat transfer from the vessel occurring during the loading and pressure come-up period. Since this method involves loading a lower temperature sample into a higher temperature vessel, the sample will quickly gain heat and increase temperature even before pressurization has started. As a result, this method requires loading the sample into the vessel according to a defined, standardized time schedule to ensure a consistent sample temperature at the time of the start of pressurization. In situations where a high radial temperature gradient develops in the sample prior to the start of pressurization, these gradients can persist after pressurization and concurrent with compression heating acting on the sample uniformly. Thus, this technique is not simple to implement.

Effects of Packaging Material on Heat Transfer in HPP

Additional factors such as pressure medium, packaging material and packaging method, and sample size contribute various heating and

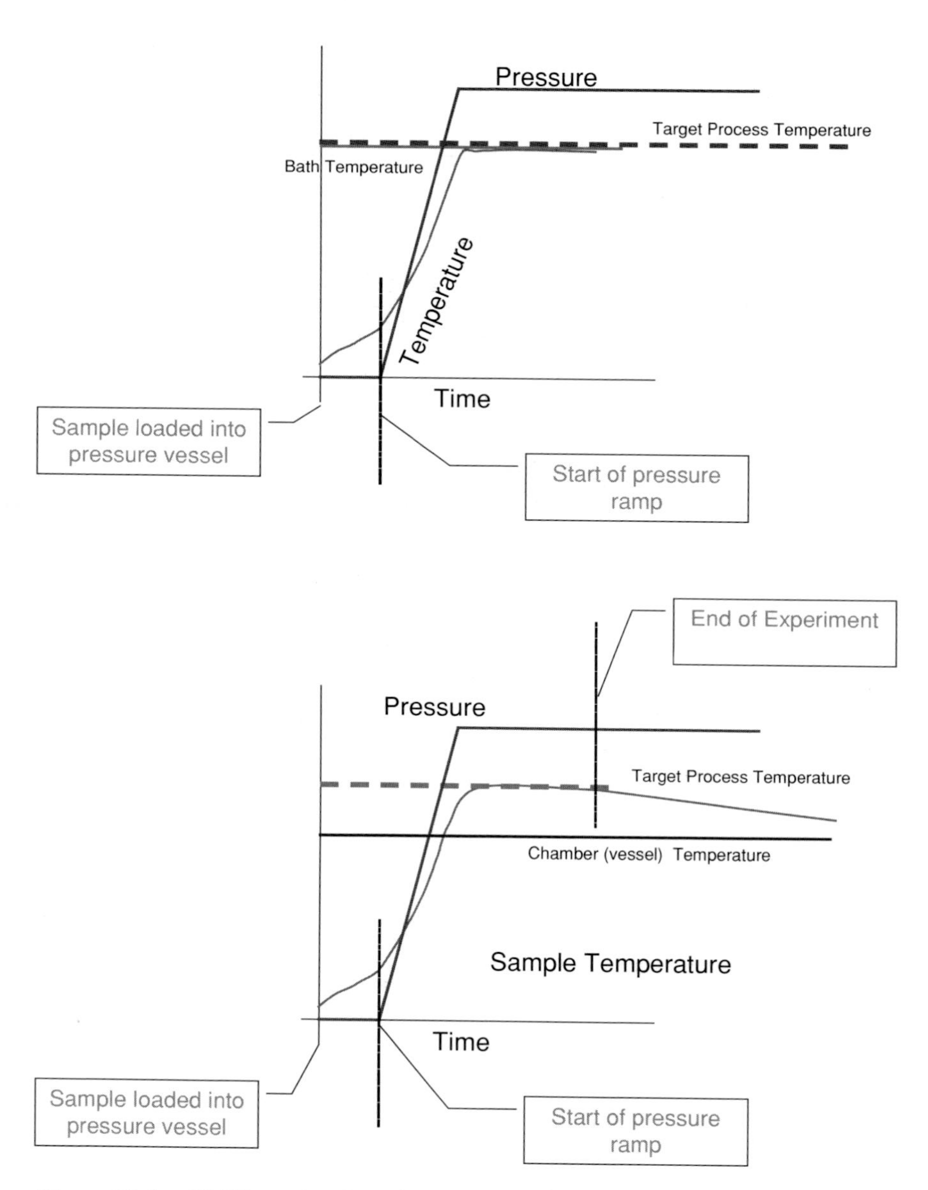

Figure 11.4. DMT requires insertion of sample into a pre-heated vessel. Pressure starts when the sample warms to a pre-set temperature. The vessel temperature can be set for the process temperature (a = long runs), or intermediate temperatures (b = enhances temperature response).

cooling interactions that complicate the overall behavior of the system and should be taken into consideration to control temperature and obtain highly reproducible results in experiments using small-scale equipment. The pressure medium should have compression heating characteristics similar to those of the sample. Water is generally the best pressure medium to use for most foods (unless the samples have particularly high oil contents) or culture media. The properties of the packaging material can have a significant effect, especially if the sample mass and the packaging mass are of the same relative magnitude. Plastics such as polyethylene have compression heating behaviors that are significantly greater than water and can transfer heat to the samples during the hold time in the process called secondary heating.

For long duration experiments, the vessel temperature is frequently set to the intended process temperature (Figure 11.4A).

Under this condition, experiments of infinite duration can, in principle, be conducted without the concern of sample temperature loss. For certain intermediate duration tests, the best procedure uses a vessel temperature that is lower than the intended process temperature. The lower vessel temperature allows overheated packaging material to be cooled and reduce the effect of sample temperature overshooting (Figure 11.4B).

Acknowledgment

This work was made possible by funding from the US Army Natick Soldier Research, Development, & Engineering Center through a Dual-Use Science and Technology (DUST) program. This DUST consortium consists of contributing partners from academia, other government agencies, and food companies, including Hormel Foods, ConAgra Foods, Kraft Food, Basic American Foods, Washington Farms, Unilever, and Baxter Healthcare.

References

Rasanayagam, V., V.M. Balasubramaniam, E. Ting, C.E. Sizer, C. Anderson, and C. Bush. 2003. Compression heating of selected fatty food substances during high pressure processing. *J. Food Science* 68(1):254–259.

Index